T0073225

SCIENTIFIC COLLABORATION

SCIENTIFIC COLLABORATION

STRATEGIES FOR SUCCESSFUL RESEARCH TEAMS

JEANNE M. FAIR

Johns Hopkins University Press
BALTIMORE

Printed in the United States of America on acid-free paper
9 8 7 6 5 4 3 2 1

Johns Hopkins University Press
2715 North Charles Street
Baltimore, Maryland 21218
www.press.jhu.edu

Library of Congress Cataloging-in-Publication data is available.

ISBN-13: 978-1-4214-4744-5 (hc)
ISBN-13: 978-1-4214-4745-2 (ebook)

A catalog record for this book is available from the British Library.

To all my research collaborators, mentors, students,
colleagues, and administrative staff,
who have worked together in the service of science,
making the world a better place

To my research colleagues in the Ukraine,
who are fighting for democracy, science,
and their freedom

CONTENTS

ACKNOWLEDGMENTS

The journey for this book began 35 years ago, with my first scientific collaboration and the experience of feeling the intense highs and lows of working with other people. I am indebted to all my collaborators, managers, and everyone else who has played a role in supporting this science. They often say that authors write their books for the "me I used to be," and that is the case for this book. Looking for resources on how I could collaborate better was how I slowly learned methods to communicate more effectively or speak up about my expectations. And it was through my failures that I learned the most valuable lessons.

My gratitude goes to all the scientists who could take the time to be interviewed and responded to my questions about what makes a successful collaboration. Almost all of these individuals were interested in sharing their insights and opinions on scientific collaborations. My research, and science in general, have benefited enormously from the researchers who have sought to better understand the science of team science. An enthusiastic high five goes to all the people who have dedicated their time and energy to investigating the practices of researchers and improving the multidisciplinary collaborations of today.

I want to give a big thank-you to the collaborators in my current projects, particularly the Avian Zoonotic Disease Network team, a partnership between Michigan State University, Jordan's Royal Scientific Society, the country of Georgia's National Center for Disease Control and Public Health, CRDF Global, and our Ukrainian colleagues. I am grateful for everything I have learned from my international partners in One Health, from Kazakhstan to Kenya. Thank you for your dedication to making the world a better and safer place.

Several researchers were especially giving of their time in talking with me about their ideas and work on collaborations and scientific networks.

I could not have completed this book without discussions with Barry Bozeman (Arizona State University), Charles Vogl (author, speaker, and advisor), Erik Olsson (Lund University), Bill Karesh (EcoHealth Alliance), Samuel Scheiner (National Science Foundation), Jason Blackburn (University of Florida), Rita Colwell (University of Maryland), Marty Stokes (Battelle Memorial Institute), Falgunee Parekh (Epipointe), Ariel Rives (University of New Mexico), Jeni Cross (Colorado State University), Jen Owen (Michigan State University), Judith Scarl (American Ornithological Society), Mac Hyman (Tulane University), and, lastly, Tim Wright, Martha Desmond, and Abby Lawson of the newly founded Avian Migration Program at New Mexico State University. I am grateful to Jim Nolan for telling me that I was the right person to write this book and to Terry Cutler-Broyles, Lorraine Rumson, and Kathleen Capels for their editing expertise.

I have been fortunate to have spent 28 years at one institution, Los Alamos National Laboratory. There I learned the most from my colleagues: Andrew Bartlow, Patrick Chain, Cyler Conrad, Helen Cui, Karen Davenport, Sara Del Valle, Alina Deshpande, Tracy Erkkila, Paul Fenimore, Jennifer Foster-Harris, Cheryl Gleasnor, Charles Hathcock, Nick Hengartner, Elizabeth Hong-Geller, Shannon Johnson, Carrie Manore, Kirsten McCabe, Benjamin McMahon, Earl Middlebrook, Harshini Mukundan, Jennifer Payne, Dina Siegel, Brent Thompson, and many more. Thanks for all the successes—and failures.

I have been most fortunate to have had great science mentors throughout my life. From my master's and PhD advisors, Patricia Kennedy and Robert Ricklefs, to the mentors who have supported me throughout the years: Bette Loiselle and John Blake (University of Florida), and Babetta Marone (Los Alamos National Laboratory). I am deeply grateful for the insights of all my students in the past, who have pushed me the most to communicate better and define expectations. Moreover, my appreciation extends to all the current and future research students who are not afraid to make things better in science with regard to inclusiveness and how we treat each other.

Lastly, heartfelt thanks to my family, who remind me often to keep working on my communication, empathy, and collaboration skills.

SCIENTIFIC COLLABORATION

INTRODUCTION

Bad collaboration is worse than no collaboration.

MORTEN HANSEN, *management theorist and author*

Beethoven wrote great music for the bass violin in his symphonies. By great bass music, I mean interesting music that includes melodious licks, not just the background beats. "Interesting" also means music that is more challenging to play. It was Beethoven's Symphony no. 7 that I was playing one day on the bass violin in my university orchestra. As much as I loved this symphony, and as hard as I had practiced this particular bass part, I was not able to focus. I was thinking about a frog that I had seen earlier that day at a pond near where I lived. The frog was not anything special from what I could tell, but I was curious about what species it was. Where did the frog come from? How did it get there? Was its breeding season over? I was sitting on my stool, struggling to keep up with the orchestra, when I decided it might be a good idea to change my undergraduate major from music.

Luckily, I was still only a freshman and had not missed many classes, which would have added years to my alternate course of study. That day I made my way over to the Biology Department and found a professor willing to talk with me and be my advisor, with one caveat: I would have to take Biology 101 the next semester and see how I did. I signed up and eagerly waited for the spring semester to start. I found the class as challenging as any Beethoven symphony bass lick, but it was also engaging and inspirational, and I knew I had found my path—in science. A path where the cobblestones represented asking questions, testing hypotheses, designing an experiment, collecting data, analyzing it, and then going back to the original question. At this point, my questions were not unique ones, but were simply related to learning everything I could from my zoology classes. Over the next three years, I took more than 15 "ology" classes: ichthyology, herpetology, ornithology, comparative endocrinology, mammalogy, parasitology, and ecology, to name but a few. I loved this new path, and I was anxious to go on to graduate school and learn to ask my own questions.

Thinking back to sitting on my stool playing Beethoven, I can also remember standing in the hallway of the Zoology Department, reading a poster on the wall advertising the School for Field Studies. It depicted a group of students looking through binoculars at a giraffe. I knew immediately that was something I wanted to do, despite the fact that I was working three jobs to afford my regular classes. Nonetheless, I sent away for a catalog, and when it arrived in the mail, I flipped through the courses available in exotic and amazing places around the world. My catalog grew dog-eared and faded as I carried it everywhere I went for the next several weeks. As appealing as the other locations and offerings looked, my heart was set on the wildlife management course in Kenya—and on that giraffe. When I received a fellowship to attend the program, words were not enough to express my excitement and gratitude.

In Kenya, I thrived on the experience. If I was awake, I was learning about new things: new animals, new ecological theories, new cultures. I also made new friends, including my best friend to this day. I had already submitted my first scientific publication, naming two new species of tapeworms from rodents in Israel (that is another story), but Africa gave me my first taste of setting up an experimental design to test a hypothesis and collect data. For that class, we were interested in whether the thorns on acacia trees deter giraffe herbivory. We selected a set of trees in areas frequented by giraffes on the ranch where we lived. Climbing ladders, we set about clipping half of each tree's thorns. On my return there the following year as a teaching assistant, I examined the trees. It turns out that giraffes *do* eat more leaves where the thorns have been clipped. But giraffes also can manage, as they have for 50 million years, to maneuver around those 3- to 4-inch thorns.

As much as I relished the science we did in Kenya and the learning environment there, this was also where I discovered that while I love research, I revel in doing it with other people. I treasure exchanging ideas, debating theory, organizing the daily plan with everyone, and finding out each person's strengths and weaknesses, as well as how they might contribute to a project. When I became a teaching assistant there, we invited Kenyan students from the University of Nairobi to come to lectures out on our ranch. It became even more exciting to share our experiences of learning between the two cultures. As much as I adored biology and research, however, I missed being part of a symphony orchestra—working together in complete harmony to create something beautiful and valuable. But I found out over the years that scientific collaborations are also like playing a symphony.

The truth is, regardless of whether you are an introvert or an extrovert, working collaboratively is the foundation for being a successful scientist in the twenty-first century. Science projects are getting larger and more multidisciplinary with each passing year, and there has been an increase in the number of teams over the last 40 years. Questions that need to be answered by science are getting to be more complex and difficult to resolve. We need researchers from many different backgrounds and disciplines, working together to tackle global challenges such as climate change, infectious and other diseases, and social issues that are dependent on both psychology and sociology, as well as to address questions about how technologies are impacting humans and global economies.

The September 2018 issue of *Science* contained a review by Martin Enserink entitled "Science under Scrutiny," which was also highlighted on the cover. It discussed individuals who now focus on how research is done and why it goes wrong. Anyone who has participated in collaborative science understands that sometimes it can go beautifully right, and sometimes it can go catastrophically wrong. In the new science revolution of multidisciplinary, fast, large-scale research, we need to ask ourselves why things go wrong in collaborative projects. And, as importantly, why things go right. How do dream collaborations emerge that not only answer important questions in science, but also add deeper value and emotional connection to our lives? This book aims to help point out what can go wrong in collaborative research and present a few strategies to help build better collaborative teams and avoid common pitfalls.

Researchers who study scientific publishing are called "journalogists." The aim of this emerging field is to study the quality and rigor of peer-reviewed journal articles and track how research findings are reported on, the quality of the experimental design and the resulting analysis, and how often, in the case of clinical findings, relevant data are left out or not published at all. The underlying message of the new work by journalogists is that if we understand more about what we're doing, we will be able to do it better. Reporting on research findings is the foundation of the scientific method. It depends on rigorous experimental design, with ample sample sizes and appropriate analyses, as well as studies building on previous work and using hypotheses as a guide in moving a field forward. With pressure for researchers to publish more and the advent of predatory journals, the system is primed to accept less than rigorous—and even downright sloppy—science. This situation can breed misrepresentations of results and data fabrication.

While publishing is critical to scientific research, we could improve many of its other aspects if we had a better understanding of them. This book, along with the work of other key investigators in the field of the "science of team science," is an attempt to understand how we do things in science and then learn how we could do them better. In this case, I will examine scientific collaboration. How, overall, are scientists working together? What makes for a great collaboration? Why do some collaborations fail? By learning from teams in the business and corporate environment, what could we do better to improve our collaborations?

In college and graduate school, we focus on building our technical skills, understanding a particular field, and thinking critically. We learn about the scientific method and, I hope, take numerous courses in statistics. We practice forming hypotheses and thinking about potential confounding factors in an experimental design. We learn to control our variables and improve our skills with high-tech instrumentation. We read article after article on our specific subject, as well as countless books, enjoying even the minutia in our field of study. In short, we become highly trained, critical-thinking scientists. We are taught to be independent. Then, after graduate school and a postdoc position, we are often, and frequently suddenly, expected to be part of a research project team, designed to bring multidisciplinary experts together to answer larger scientific questions. Sometimes working with the team is rewarding, fun, and productive. At other times, the team itself slows things down and puts glue in the research engine.

What goes wrong? What goes right? Given that we've been taught to become independent researchers, we may have less experience in working with other humans, with the key word being "humans." No longer are controlling for variables and coding in R for some novel statistical analysis the most important factors in moving a project forward. When working with other humans, the most important thing may be understanding personalities and communication styles, building and losing trust, and comprehending our own and sometimes others' puzzling emotional reactions to our colleagues, as well as their reactions to us.

Thankfully, many people around the world have been studying how teams of people work best and examining what can go wrong. As a close friend of mine says about every problem or challenge, "There's a manual for that. On YouTube." Yet, when it comes to scientific collaborations, what we often fail to consider are the feelings and thoughts that crop up when working closely with humans who are not our spouse, our family members, or a few close friends. Science and emotions don't usually go together—unless you are a researcher on the psychology of emotions. If we promise to give the data to our colleagues the next day, but our child gets sick and keeps us up most of the night, so we don't get the data to the team, what do we feel? Maybe vulnerability and shame? And how do we respond and behave if we are unaware of our feelings? We may lash out or pull away from our colleagues.

If peer-reviewed publishing is a foundation for moving scientific fields forward, then trust is the basis for moving scientific collaborations forward. Trust, however, is multifaceted and not all that simple. As Stephen M. Covey said, "Trust is the one thing that changes everything." So how do we build trust and keep it? How do we develop mindfulness about employing trusting behaviors in our everyday life that can help propel our science teams to success? Knowing the technical aspects of research is not enough. Scientists must begin to hone their skills not only of working within the same space, but also of working collaboratively. With the competition for science funds becoming more rigorous with each passing year, researchers must focus on maintaining a competitive edge with others and achieving peak performances.

But what is "peak performance" for a scientist? It may mean being the most well-versed person in the room on a specific topic within a particular field. For example, in ornithology, your area of expertise may be the impacts of food limitation on immunology for a particular taxon group of wild birds during the breeding season. Keeping up with the relevant literature is another challenge. With thousands of journals, ranging from broad topics to extremely focused ones, it is difficult enough just to keep up with the titles of intriguing articles, let alone read and digest them, as we are expected to do. The other aspect of science that is changing rapidly is technology, from both a computer-based and an analytical standpoint, as well as through new measurement techniques. Researchers must keep up with changing statistical methodologies and computer code in order to correctly analyze their data without incorrectly inserting any statistical assumptions along the way.

Peak performance for a researcher also means working successfully with other researchers, some of whom are from diverse backgrounds and situations, and who are sometimes in a good mood and sometimes in a bad mood. They may or may not be introverts, or they may feel entitled at times because they have published papers in *Science* and *Nature*. They may be going through a divorce, or perhaps their mother recently died of cancer. They may be struggling to learn the newest statistical software, which has changed since they went to graduate school, while also carrying a full class load and juggling both a new National Science Foundation project and 6-month-old twins. They may not be good at communication because they did not learn effective and efficient ways to carry out crucial conversations. Whatever the reasons, our job is to find empathy for our colleagues and understand that we are all doing the best we can with the resources available to us. This is a first step toward better collaboration.

Not only is how we *collaborate* as scientists changing, but how we *communicate* our research with other researchers and with the world at large is altering at lightning speed. Young researchers are now using social media to release their findings and establish a laboratory "brand." In 2019, before the COVID-19 pandemic, I participated in my first Twitter "conference" for a scientific society. During the pandemic, researchers had been attending scientific conferences virtually. How has this new virtual space changed collaborations and ways of sharing our information?

The good news is that many researchers in the areas of psychology and business have done their own phenomenal studies on what makes exceptional teams tick, and they have provided insight into how to have better collaborations. Some researchers, such as Dr. Barry Bozeman at Arizona State University, have focused their careers on finding out why scientists collaborate and what makes for a dream collaboration. It is through the efforts of scientists who study team science, as well decades of work done on successful business collaborations, that we can have a road map for learning how to fruitfully collaborate. This ultimately leads to better—and potentially transformative—science, as well as higher-quality publications, more funded proposals, and increased citations of the work. The greater the number of citations a publication has generally means that the work has a greater value for the field and for society in general. Even when science is not purely experimental, but instead is more about development, theory, or modeling, the principles for successful collaboration are the same.

How do you define success in your own science career? While the above describes peak performance in science, the truest definition of success is enjoying both the end result and the process of getting there. Another sign of peak performance is feeling connected to others and enjoying the laughter, gifts, and idiosyncrasies of our colleagues. Sharing our gifts and "leaning in" (persevering in spite of risk or difficulty) to learning with others who are following their own paths can lead to true joy and contentment. Conversations between scientists can now contain words such as "multidimensional spectra analytics" and "compassion" at the same time. Learning how to effectively communicate, to be trustworthy and build trust, to be compassionate and have empathy, and to be technically competent is the goal of this book. Through a combination of stories and resources, it will identify the cornerstones for successful collaboration, as well as help people operate at a peak performance level and work together to create engaging, fun, and sometimes transformative relationships. Often it only takes a few little things to metamorphose an average collaboration into a dream team.

CHAPTER 1
TRANSFORMATIVE COLLABORATIONS

> When you wait until the day of the outbreak to exchange business cards, the pathogen has already won.
>
> GARY SIMPSON, *former medical director for infectious diseases, New Mexico Department of Health*

Can you remember 1993? There are the rare people who can clearly recall what they were wearing on June 23, 1993, when they saw *Jurassic Park* at the movie theater. The rest of us mostly recollect years by major events, milestones, or changes in our own lives. A friend of mine helped me go back in time by lending me the DVD for CNN's *100 Defining Moments of the Year* for 1993. It was like stepping into a time capsule. Among many other events, Clint Eastwood's movie *Unforgiven*, set in 1880 in Wyoming, won the Academy Award for Best Picture, while another real-life story of life and death unfolded in the Southwest.

A HEALTH CRISIS IN NEW MEXICO

Gallup, New Mexico, is a generally sleepy town off Interstate 40 (previously Route 66). The community is around 48% Native American (Navajo, Hopi, and Zuni tribes), according to the 2020 census. Gallup has strong ties to the railroad, which helped establish the frontier town in 1881, and it is one of the only towns to have successfully fought against putting their 800 Japanese residents into internment camps during World War II. Gallup is the county seat for McKinley County, and on May 14, 1993, the courthouse received a call that would initiate one of the biggest scientific mysteries of the twentieth century. It would make CNN's list of the most defining moments of 1993.

THE BEGINNING

Dr. Richard Malone, the medical investigator for McKinley County, took the call. It was about a 19-year-old man named Merrill Bahe, who had been brought to the Gallup hospital and died shortly after arrival. When Malone followed up with Bahe's family to ask what had happened, they responded that they had been on their way to a funeral, and he had simply collapsed. "Whose funeral were you going to?" Malone asked. "His fiancée's, Florena. She died five days ago, after she similarly collapsed." This response probably sent chills up and down Malone's back. Bahe's funeral was scheduled to start 15 minutes later, and it would be very difficult for Malone to get any additional information. The young couple was Navajo, and, according to tribal tradition, you did not speak of the dead for at least four days.

The next calls by Malone were to the New Mexico Department of Health, turning over the case to them, and to his friend and colleague, Dr. Bruce Tempest, the director of the Indian Health Service, a federal agency. Tempest was the first physician to look at the symptoms of and laboratory results for the young victims. He told Malone that these findings contained unfamiliar characteristics, though they were similar to a few other recent cases that had come to his attention. Tempest called all the New Mexico hospitals with similar cases and found a total of five people who had recently died in essentially the same way: pulmonary edema, where their lungs filled up with fluids. This now had all the hallmarks of an epidemic. Dr. Gary Simpson, the medical director of infectious diseases at the New Mexico Department of Health, was the next person Tempest called, and a small team was formed. Later that same day, two more cases were reported. Both Tempest and Simpson ruled out common infectious diseases in the region, which included plague, inhalation anthrax, and tularemia. At this point all they knew was that this unidentified disease had a 100% mortality rate. Other top health officials were brought in to brainstorm about what it could be, and they came up with three possibilities: a mutated form of influenza, a hemorrhagic fever, or a new disease.

The team continued to grow over the next few days as epidemiologists were brought in to interview the families. The New Mexico Department of

Health sent letters to all the doctors in New Mexico to be on the lookout for both new and old cases. As with most infectious diseases, the first indications were "flu-like symptoms," which, in the current disease, were followed by a misdiagnosis. By May 27, less than two weeks after Bahe's death, there were 16 new cases. Dr. Mack Sewell, the state epidemiologist at the Department of Health, knew it was time to call the national experts at the Centers for Disease Control and Prevention (CDC) in Atlanta. On May 28, the CDC's Special Pathogens unit was brought into the investigation. Within three days, blood and tissue samples started arriving at the CDC from New Mexico. Dr. C. J. Peters, the chief of the Special Pathogens unit, put together a team. A CDC outbreak investigation team typically comprises two to three people, but in Peters's words that day, he "knew this needed a full court press," so initially a team of four people was sent to New Mexico.

On Memorial Day weekend in 1993, two days before the CDC team's arrival, patients with similar symptoms were starting to show up at the University of New Mexico's medical center in Albuquerque. They were placed into separate, single-person rooms in the hospital. These rooms were then sealed, and the only access was from an isolation room. When such patients were brought in, they were immediately stabilized, but x-rays showed their lungs filling with liquid in less than four hours. The CDC team arrived on May 29. In the initial meeting between the New Mexico team and the CDC experts, they were given the news that a little girl who was suspected of having the unknown new disease had died. Simpson, now head of the New Mexico Task Force, was deeply troubled and pained by this news about the little girl—and the growing epidemic.

By this time the media was reporting on the disease outbreak, unfortunately named the "Navajo Flu." This conjured up negative stereotypes and added stress to the impacted communities. Navajo people were denied service in restaurants, and other types of discrimination took place. Fear of and ignorance about the disease led to a Navajo couple being denied entry into Canada at the border. In response to being treated poorly and ostracized, the Navajo people retreated into their communities and refused to talk to outsiders. While understandable, this led to challenges with the team's investigation. On June 3, 1993, a healing ceremony lasting five hours was held by 300 medicine men to pray for the victims of the epidemic. By this time, the disease had spread to Utah, Nevada, and Texas.

With pneumonia, it usually takes from three to five days before the lungs start filling with fluid. In the 1993 New Mexico victims, the same process was taking three *hours*. But if patients could be stabilized prior to pulmonary edema occurring, they could be saved. Still, the mortality rate was an astonishing 77%. As director of the task force, Simpson was holding two press conferences a day. He remembers the situation as both overwhelming and terribly depressing.

The original young couple, Merrill and Florena, had a 6-month-old infant who became ill soon after they died. The baby, named Maurice, was placed in quarantine at the hospital, but after a few days it became apparent that he did not have the same disease as his parents, and he soon recovered. Tragedy was not finished for this family, however. Merrill's brother-in-law moved into the now-vacant trailer, and a week later, he came down with the disease, as did his wife. This provided a clue: if four people in the same household were infected, it meant the disease was contagious. Or so the disease investigative team thought.

AT THE CDC

Back in Atlanta, the CDC began their research with serology tests, in which known pathogenic microbes are put into small wells, and the serum from the blood of the victims is added. If a victim's serum contained antibodies to the pathogen, meaning that person had previous exposure to the microbe, the reaction between the antibody and microbe could be measured. This process continued for several days, with all samples testing negative for antibodies to known pathogens. Finally, there was an antibody reaction to a type of virus known from Southeast Asia, but not North America. While this was not confirmation of what was going on, it was a foot in the door. That victim's antibody seemed to have come from an infection most closely related to a pathogen called a hantavirus, from Southeast Asia. At the time, the CDC knew of approximately 20 strains of hantaviruses around the world, with 5 or 6 of them being deadly. All of them were found in rodents.

Hantaviruses were first discovered in the 1950s and named for the Hantan River in Korea, where, during the Korean War, 3,000 United Nations soldiers were infected. The disease in Korea, however, affected the kidneys. So while the serology tests seemed to get a hit on a hantavirus from Korea, it still did not add up. How had the virus gotten to New Mexico, and why did it impact the lungs but not the kidneys?

Next, the Molecular Biology unit of the CDC, led by Dr. Stuart Nichols, took over the investigation of the positive samples. By then, a new molecular technique had been developed, called the polymerase chain reaction, or PCR, which allowed sections of genetic amino acid code to be amplified, in order to make the amino acid composition of the sequence visible. This test looks for specific sequences found only in certain viruses or bacteria, in order to confirm their actual presence in a sample. Due to the relative newness of this technique, it had not yet been tried in real world instances. This was to be its first application in pursuit of identifying the cause of an outbreak. Working around the clock, in the early morning hours the researchers detected the genetic "fingerprint" of a hantavirus. But it was not a known version of the virus. It was a brand new one. The team left a note for Nichols to find when he got in that morning. It took a total of 19 days from discovering the first victim to identifying this new virus. The news was communicated to the world on Friday evening, June 6.

Once the CDC team confirmed that the new disease was a hantavirus, they knew there was no person-to-person transmission, and that it was probably carried by rodents in their urine and droppings. Their focus then shifted to capturing small wild mammals in the areas near the victims, an effort that started on June 7. While several small-mammal species were found to harbor the virus, the primary carrier was the deer mouse, or *Peromyscus maniculatus*. It is an adorable mouse, with large eyes and ears, that lives all across North America. At this point, Drs. Bob Parmenter and Terry Yates joined the task force team. Both doctors were professors of biology at the University of New Mexico (UNM) and studied small mammals in the Southwest. Parmenter had several long-term sites in New Mexico, from which he had detailed population and environmental data on deer mice for over 15 years. Yates, a neighbor of Simpson's, was the director of the Museum of Southwest Biology at UNM. He, too, studied deer mouse populations in the wild. Both doctors were brought in to start capturing small mammals and taking samples.

Back at the CDC, researchers took the samples into the highest-level containment laboratories available, a biosafety level 4 sealed room. Here they infected mice with the samples in order to answer some key questions. All work shifted from identifying the virus to learning how to culture it in tissues and then isolate it. The next question to answer was how long mice carry the virus, which would answer how long they are contagious and how, specifically, the virus is transmitted.

PROGRESS IN NEW MEXICO

Staff at the University of New Mexico Hospital continued to use known methods on all its new patients with this disease, and they also began working on other diagnostics and treatments. Incoming patients were treated with the antiviral medication Ribavirin, which turned out to not be effective. Dr. Brian Hjelle, a professor at UNM, developed a rapid test for the virus, which took only 24 hours to arrive at results, but patients continued to die within four hours after being admitted. Medical teams found that if patients could get to a hospital and be stabilized before the final pulmonary edema effects set in, morality rates were reduced. They also found that recovery was as rapid as the patient's initial decline with the disease—as little as three to four days.

In the landscape of sagebrush, pinyon pine trees, and chamisa shrubs around the outbreak area, the field team continued sampling small mammals. Parmenter pored over his long-term data on environmental conditions and noticed that small-mammal populations exploded after a wet year. Higher amounts of precipitation led to greater plant and insect productivity, and that helped increase small-mammal populations the following year. Epidemiologists talking with the Navajo people learned that in both 1918 and 1933, there were similar outbreaks in the region, and those years followed wet ones. The Navajo medicine men told of how they knew to stay away from mice and not "let them run across your clothes." The hypothesis by the mammalogists, epidemiologists, and medicine men was that an explosion of mice would lead to more contact with humans and, thus, transmission of the disease. Deer mice populations in New Mexico were peaking in the spring of 1993.

The next big question to be answered was how deer mice got the virus. Had the virus always been in the region, or was it introduced into the Southwest from someplace else? As one conspiracy theory had it, maybe the mysterious nuclear weapons laboratory in the region, Los Alamos National Laboratory, had released the disease either by accident or for some diabolical reason. Or maybe it came from Fort Wingate, a military installation near Gallup. Simpson talked to his neighbor Yates about these questions.

Picture them meeting at the end of a dirt road and having a conversation about the disease along the following lines. Yates says enthusiastically, "Well, let's find out if it has been here a long time!" "How can we do that?" Simpson asks. Yates then tells him about the thousands of small-mammal specimens he has, going back 50 years, at the Museum of Southwest Biology. They could test the archived specimens for the disease and at least discover how long it had existed in the area—and get Los Alamos off the hook. Yates probably immediately went back to the museum to identify specimens and collect samples to look for the presence of the new hantavirus.

Fortunately, the natural history collection at Texas Tech University (TTU)—where Dr. Robert Baker, a fellow mammalogy professor, worked—also had specimens. With no previous outbreaks, and with the virus remaining unidentified until this point, the medical community was skeptical that they could have missed it being endemic to the region. How could it have escaped notice? With the virus now infecting numerous people, there were only three possible options. First, it could be the result of coevolutionary and host-specific processes that allowed the virus to be present in the small mammals all along. Second, maybe it was a rapidly evolving virus that had recently adapted its makeup, due to climate change. Third, it could have been newly introduced. Yates and Baker completed their testing of deer mice specimens from UNM and TTU, which showed that the outbreak was not the result of a recent introduction. It had been in the deer mice population in the Southwest all along, going back decades. Once the investigators knew what they were looking for, they unearthed studies that documented human cases of deaths from hantavirus in New Mexico in 1958. Moreover, as noted before, the oral traditions of the Navajo people demonstrated a longtime understanding about the dangers of deer mice in their homes.

By the fall of 1993, the cases of hantavirus were dropping off, and the task force team members' lives were returning to normal. In the end, the outbreak in 1993 infected 115 people, killing 59 of them. The CDC first named the disease the "Muertos Canyon hantavirus," after the area near the first cases. After concerns were voiced about that appellation, it was renamed "sin nombre," meaning "no name." After the CDC investigators left New Mexico and returned to Atlanta, the New Mexico and Texas group members stayed in close contact almost daily. This is where the story of the transformational science of that team begins.

THE RAMBO TEAM

Once the epidemic itself was contained, the next question to tackle was what were the specific drivers for the current 1993 outbreak. The New Mexico and Texas team brought in climate scientists, in order to understand the dynamics between the environment and small-mammal populations. Using data collected from Sevilleta Research Station, a long-term ecological research site managed by UNM, the team learned that high mouse populations and hantavirus prevalence were linked to abundant seed production after substantial rains, and variations in the deer mouse population were linked to the El Niño / Southern Oscillation, or ENSO. In El Niño years, the Southwest typically has more precipitation than in other years. This includes higher levels of snowfall in the winter and wetter springs, directly leading to an increase in plant productivity, including seeds. During their breeding season, deer mice feed on these seeds, which are rich in calories, increasing both their litter sizes and survival rates. Since their predators do not reproduce as quickly as the mice, there is a lag time before predators such as bobcats, snakes, owls, and raptors can affect the small-mammal populations and keep them in check. The climatologists showed the presence of El Niño conditions throughout 1991 and 1992.

In rural areas, enclosed structures, such as barns and utility sheds, allow infected mouse feces to accumulate in the dirt. The contaminated dust is subsequently disturbed by wind or other mechanisms, such as sweeping, thereby exposing unsuspecting persons to particles that can be inhaled. In one of the hantavirus cases, vacuuming was the cause. This method of contracting a disease through indirect exposure, which the team discovered, was somewhat unusual, because most infectious diseases are spread through direct contact with a biotic vector (animal, human, or insect).

Previous assumptions that a biotic vector was required to contract a disease after exposure to it had hampered the ability of researchers (e.g., Paul Zeitz and coauthors in 1995) to conceive of possible mechanistic solutions to the problem, as well as limiting the ability of social science researchers and public health officials to create effective prevention programs. One such effort, the disinfection of indoor areas with a 10% bleach-and-water solution or Lysol before sweeping, is the best known preventative measure. Almost 30 years after the outbreak, most New Mexicans now know to take precautions around mice droppings. These safeguards have undoubtedly reduced the number of infections over the years.

The scientists in New Mexico and Texas began putting the pieces together and publishing their results. They spent a lot of time together, both inside and outside of their workdays, poring over the data and telling their story. Much of their time was spent processing data on the outbreak and the disease's impacts on the victims, their families, and the region. The increasingly close-knit team developed an informal network they called RAMBO, which stood for the Research Association for Medical and Biological Organizations. Spearheaded by Simpson, his neighbor Yates, and Baker from Texas, the RAMBO team often met in the evenings at their houses to talk about their data and present new ideas to each other. This was a group I was honored to be a part of many years later. It met regularly for 25 years.

The hantavirus outbreak team considered themselves lucky to have figured out the puzzle pieces of this unknown disease so quickly, and they attributed much of it to serendipity. The right people to solve the next steps of the mystery had suddenly appeared at opportune moments. The relevant technologies were advanced enough to help dig up the next clue. Although they felt serendipity was at work—and indeed it may have been—this team was an extraordinarily apt example of transformative research.

TRANSFORMATIVE RESEARCH

The idea of transformative research was first developed by Thomas Kuhn in 1962, in his classic work on the structure of scientific revolutions. According to the National Science Foundation, transformative research is "driven by ideas that have the potential to radically change our understanding of an important existing scientific or engineering concept or leading to the creation of a new paradigm or field of science or engineering. Such research also is characterized by its challenge to current understanding or its pathway to new frontiers." While both crisis and fortuitous circumstances were the catalyst for this group of scientists and public health professionals to come together regarding hantavirus, the team itself knew that a central tenet of the group was to understand how transdisciplinary scientific collectives can be organized more effectively.

When we think of teams in terms similar to those we use for individuals, we can apply some classic ideas about psychology and human potential. One is the hierarchy of needs, developed by psychologist Abraham Maslow.

He noted that individual needs fall into categories that build on each other: physiological, safety, love and belonging, esteem, and self-actualization (self-fulfillment). Maslow believed there were hierarchical dependencies between these. For example, a person could not achieve self-esteem unless basic physiological, safety, and love and belonging needs were met. This theory of human behavior is sometimes contested and competes with many other theories. Nevertheless, for our purpose we can think of analogous categories for team needs. In the hantavirus outbreak situation, physiological requirements corresponded to the foundational elements on which a team conducts research. In a 2013 paper published in *BioScience*, I and a few of my colleagues in New Mexico, including Simpson, outlined how the hantavirus outbreak team was an example of transformative research, as well as a good example of the hierarchy of team needs. We lumped safety, social, and esteem requirements into a category called "mutualism" (trust, friendship, respect for others), and we created a new category called "integration," which represents the synthesis of different data, knowledge, and perspectives into creative new information structures, generating co-emergent ideas.

Many elements were needed to already be in place in order for the serendipitous encounters in the hantavirus outbreak investigation to occur. First, each researcher had invested a portion of their lives in becoming experts in their own fields. Without the deep knowledge accrued over decades, the hantavirus team could have not given the best educated guesses that narrowed their focus to the final three possibilities on their list. They would not have postulated that short-term mammal research could answer key questions about the epidemic. Deep knowledge in a particular area of expertise is a fundamental building block on which transdisciplinary research rests.

The second required element involved connections through scientific networks. These links were critical for transferring ideas, knowledge, data, and samples relevant to the outbreak's investigation. Therefore, the human requirements of transdisciplinary and transformative research are team members with both individual expertise and social networks that can be used for reaching out to other experts and connecting them with the team.

In addition, there are physical requirements in science, which are also important in achieving transformative outcomes and team success. In the hantavirus outbreak, tissue samples from the existing collections of small mammals were critical. These mammal specimens were originally collected for many reasons, and data from them were easily applied to the question of whether the hantavirus was new in the region or had been around for much longer. Scientists tend to be hoarders when it comes to data, samples, and basic information. Both data and samples collected over time represent a major investment of resources and effort, resulting in potentially reusable resources. This has been a central point of discussion during nearly every disease outbreak, from Ebola in West Africa in 2015 to the most recent SARS-CoV-2 coronavirus, which caused the COVID-19 pandemic that began in 2020. I would be remiss here if I did not point out that reusability depends on good metadata about the original data and samples. In the case of the small-mammal samples being used by the Museum of Southwest Biology for the hantavirus outbreak, metadata regarding the year and the location where they were collected were the most valuable kinds of information. The sustained curation of data and samples in a usable form can be critical in using history to solve present-day science mysteries. Retaining those collections of samples should not be a problem. But, as a manager of scientists, I have also seen how much data and how many samples are left behind after some scientists retire, with no organized archive or owner designated for them after the original collectors leave.

COLLABORATION

In our 2013 paper, my colleagues and I used the phrase "collaborative mutualism" to describe the dependencies between individuals in research teams. "Biologic mutualism," which the scientific community understands relatively well, is defined as a species' or an individual's interactions where both the actor and the recipient derive a biological fitness benefit. "Fitness" means the ability to survive and reproduce. At the individual level, it depends on both genes and environment. Like fitness within a population, transdisciplinary teams are composed of individuals with different disciplinary types (genotypes), confronting some problem context (environment), and working together to survive in the difficult research arena by producing innovative results.

The irony of science is that there are inherent tensions between competition and collaboration in research. In both human society and in nature, competition occurs when individuals vie for the same resource or territory, or for recognition. In the case of research, investigators compete for funding resources, community recognition, or even laboratory space within their own institution. Science as a whole, however, is a far more cooperative endeavor, with different groups passing along how science in their field is conducted. Such communities can collaborate by sharing data and information, reviewing theories, and constructing or evolving new ones. This, in turn, drives increased cooperation, since lone individuals cannot compete as well unless they engage with communities at multiple scales. The interaction between individuals within groups can then generate new knowledge. Like evolutionary fitness in nature, the diversity of the various participants in a community leads to a competitive advantage. Different backgrounds can produce more creative ideas for solving problems and addressing scientific challenges.

Within a research team, individuals choose to participate because they believe cooperation will lead to an enhanced ability to compete for funding and will generate findings that have a significant impact within their own discipline. If they do not believe this, they may choose to invest their time in other things, sometimes even if funding and recognition may be present in whatever they turned away from.

Before collaborating, individuals make both conscious and unconscious comparisons between the perceived potential benefit of funding or recognition and the perceived costs in time, effort, and resources. In the hantavirus team example, societal importance was a motivating factor that overrode any other issues. Young people in the Four Corners region of the Southwest were dying, and this group of scientists was positioned to change the outcome of the disease outbreak. The team did not initially think their findings would be transformative and would gain them additional funding and recognition, though these results did occur. They had all the funding resources they needed at the moment, their various home organizations allowed them to dedicate time to the investigation, the benefit to society was clear, and the costs of collaboration were reduced. Thus the decision to participate in the team was relatively straightforward. This was also quintessential science, resembling a giant puzzle. A desire to "figure it out" or "put the puzzle together" was a contributing driving force.

Studies of current early-career researchers (ECRs) have indicated that in spite of short-term career setbacks, these individuals believe that engagement in interdisciplinary efforts will better position them in the long term, as well as enable them to address issues of significant scientific and societal importance. One good example of such a perspective is a 2019 paper by Pannell and colleagues, where the authors pointed out the importance of interdisciplinary research in ecology and sought to envisage a future where barriers to collaboration were removed. ECRs who have participated in successful interdisciplinary teams have consistently emphasized the value of such teams, both for tackling complex research problems and for their own career growth. Yet many barriers remain for ECRs, which are related to the competitiveness of science and multidisciplinary research. A team of ECRs in Norway compiled a comprehensive list of these barriers in a paper on coastal research in their country. Hence we hopefully appear to be transitioning from a culture in science that is dominated by individual competition and within-discipline community cooperation to one that also balances individual competitive drives with cross-disciplinary cooperative goals. This is being facilitated in many ways. Government agencies have increased funding opportunities for inter- and transdisciplinary research. Professional journals have emerged that target inter- and transdisciplinary articles. Policies are being generated that demand research having a broad impact on society. On the other hand, collaboration is also being inhibited by organizational and institutional policies, procedures, and structures that favor individual competition within disciplines over cross-disciplinary engagement.

At the onset of the hantavirus outbreak, no single person had any answers. As the crisis evolved, everyone made observations on the fly. Team members were thrust into a situation where they had to rapidly share information and grasp each other's perspectives in order to understand how different things were related, even though those perspectives might have been outside their own disciplines. They were able to learn about various issues and knowledge gaps from their colleagues' perspectives, and then seek approaches that could address those issues. This required listening to other team members and wading through the difficulties inherent in transdisciplinary collaboration. One of the largest barriers to such collaboration involves differences in vocabulary, methodology, and epistemology. Due to the health crisis at hand, the hantavirus outbreak created an environment where the team members' conceptual differences were openly discussed. The team often used lists, diagrams, and brainstorming sessions that included all members. The only thing required to be part of the team was an open mind.

CREATIVITY

The idea of transformative learning across multiple disciplinary perspectives leads to the notion of a conceptual zone of collective research alignment that is creative, because it effectively mixes the disciplines involved. Researchers with multiple perspectives can learn about others' viewpoints and allow their own perspectives to be altered by such interactions. Their changed perspectives then enable them to expand their breadth of knowledge in their own discipline, develop mutualistic collaborations with their colleagues from other disciplines, and generate integrative, synthetic outcomes in multiple disciplines. The team of authors in our 2013 paper called this endpoint the "Yatesian Zone," in honor of our RAMBO network colleague, one of the core team members during the hantavirus outbreak. Yates, who died of cancer in 2007, was a visionary who operated consistently in this zone, both in his own transdisciplinary collaborative research and as a mentor to others. As an example, he was wonderful at pulling together intellectually diverse people and organizations to work collaboratively for the benefit of all. "He was one of the most friendly and enjoyable people who I have ever met," one of his colleagues said, echoing a common theme among the RAMBO network.

Reaching the Yatesian Zone depends not only on transformational learning, but also on the breadth of the disciplinary foundations of those involved (figure 1). It is rare for individuals from two separate disciplines to meet and have similar languages, observations, data, and methods that they can use to immediately collaborate on complex transdisciplinary hypotheses. Rather, as part of the transformational learning process, each investigator typically must "reboot" their own concepts, data, and methods in this new context. The collective focus of the group is a moving target, which emerges from transformative learning, reconceptualization, and re-observation, based on the new ideas generated. This describes both the process of knowledge synthesis across disciplines and the generation of synergistic outcomes. Synthesis efforts may require substantial time to collectively have research evolve toward this new, synergistic area—yet when it occurs, it is likely to be transformative. Integration and synthesis across disciplines leads to cognitive struggles and requires mental flexibility. While this may be difficult, it will change the way an individual views their research questions, and, as such, provide a mechanism for generating transformative science.

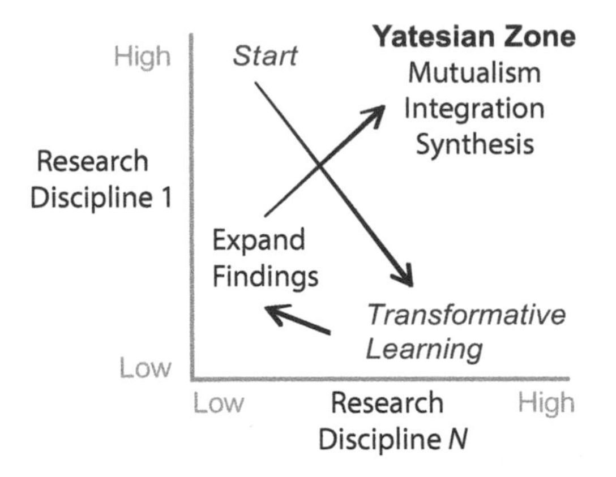

FIGURE 1. Yatesian Zone of transdisciplinary and co-emergent innovation. Arrows represent the path of researchers who begin working in the frontiers of their own discipline, engage in transformational learning in one to *N* other disciplines, expand the conceptual, data, specimen, and network foundations of their own discipline, and then are able to effectively integrate their research with others from disciplines 1 to *N* in the Yatesian Zone. This process, which may be undertaken by individuals exploring other disciplines or by collective groups, results in knowledge synthesis. *Used with permission from Oxford University Press*

In our 2013 paper, my colleagues and I suggested a direct link between transformational learning and transformative science. Thomas Kuhn, one of the great philosophers of science, highlighted the role of accumulating results or data outliers in the emergence of radically new conceptualizations in science. His theories and his numerous examples of major changes in scientific theory found that science progresses through long periods of what he called "normal science," during which incremental innovation occurs around accepted scientific conceptualizations. Data that do not fit into this mold accumulate, but they are written off as data collection errors. Eventually, anomalies become so extensive that someone finally rethinks the accepted norm, proposes a completely new conceptualization, and, after a major confrontation with those holding on to the entrenched ideas, a radical theoretical change occurs. Kuhn called this "revolutionary science."

Kuhn's model described transformational learning as being generated through a disruptive or disorientating challenge, which leads to critical discourse and reflective communication. My colleagues and I postulated a more general model, incorporating the role of disorienting data, as well as collaborative and integration dilemmas. This was exemplified in the hantavirus public health crisis and added to by the plethora of studies conducted under the emerging field of team science.

Those who aspire to participate in transformative science should begin by intentionally placing themselves within a disorienting dilemma. Since observational anomalies and crisis contexts cannot be created at will, the best known path toward purposeful transformative science is through exposure to different disciplinary perspectives and transdisciplinary contexts.

ESTABLISHING STRONG AND LASTING RELATIONSHIPS

Not only did the hantavirus outbreak team forge strong friendships, built on a foundation of trust and respect for each other, but the collaborators consciously worked to strengthen those relationships. The outbreak team members, who had stayed in touch after 1993, reconvened when the Health Information Alliance of New Mexico initiated a retreat in April 1996 to "bring together diverse organizations, associations, commissions, national laboratories, university research groups, and state offices to identify databases and other sources of information within the state that relate to human health." From the initial retreat, a pilot project was proposed to develop a model for sharing data between diverse groups that would address any community, regional, or state health problem.

Even before big hairy audacious goals (BHAGs) became popular in the goal-setting world, the RAMBO group created BHAG mission goals. The RAMBO members called them "big hairy-assed goals," although most people now call them "audacious." The RAMBO team defined the attributes they wanted for the group:

Opportunity driven (with vision)
Access to interorganizational resources
Collaborative use of diverse databases and information resources
Multidisciplinary approach to complex problems
Specialized technical resources (e.g., communications, data access and security, patient confidentiality, law)
A learning organization with processes for educating the group

From the beginning, the group met regularly, most of the time after hours and at their homes. They discussed the science of human health and infectious diseases, which included talking about wildlife, agricultural animals, humans, and the environment. Their experience with hantavirus led this team to know the importance of a One Health approach to emerging infectious diseases. The term "One Medicine" has been around for a long time, going back to the Hippocratic treatise *Airs, Waters, and Places*, written in 400 BC. Many centuries later, in 1964, Dr. Calvin Schwabe used "One Medicine" in his book *Veterinary Medicine and Human Health*. Schwabe was one of the founding faculty members of the medical school at the University of California–Davis, as well as an epidemiology professor at the veterinary school there. Hantavirus may be one of the best examples of how climate, the environment, wildlife, and humans are all integral to searching for outbreaks of infections. One Health is now more commonly used to represent the interactions of zoonotic diseases within the environment, animal and wildlife hosts, vectors, and humans.

Since the RAMBO team consisted of members from all different types of organizations, they knew they were an adhocracy. The word "adhocracy" was popularized by Alvin Toffler in the 1970s and refers to a flexible and adaptable organizational structure without bureaucratic policies and procedures. RAMBO participants followed and studied his work, and they sought to foster the characteristics of adhocracies, including being a diverse multidisciplinary group with an emphasis on expertise. They wanted an organic structure that was flexible and responsive, with no single concentration of power—hence their informal gatherings. They created a matrix, joining functionality with project-driven team building. In other words, they talked about and sought to foster a creative environment for complex, dynamic problem solving, in order to encourage sophisticated innovation.

In 1999, three years after the first retreat, the group had a defined list of words they would use to describe RAMBO. These were:

Relationships Trust Warm Friendly
Sharing Exciting Barrier-breaking Talking
Diverse Non-territorial Interactive Organic
Innovation Multidisciplinary Collaborative Earthy
Creativity Substantive Technologically Aware Curiosity

Learning Environment Open Energy Visionary
Dynamic Synergistic Productive Fun

I don't know about you, but I can't think of anyone who would not want to be part of a group describing themselves in these terms. As a model for their success, the RAMBO group chose a quote from Charles Handy, author of *The Age of Paradox*: "The secret to new growth is to start a new sigmoid curve before the first one peters out." The group studied what made traditional organizations successful, and they saw themselves as "RAMBOesque." They were also eager to share what they were learning. While most of what they discussed when they got together for dinners around Albuquerque, New Mexico, were infectious diseases, biology, public health, and ecology, they also talked about what makes organizations successful. Since the initial relationships were forged during the hantavirus outbreak, the group could see that it might take a crisis to propel society out of its mechanistic orbit. In the words of William James in 1906, later used by Jimmy Carter in 1977, "a culture may require the moral equivalent of war" to create the conditions necessary for significant and sustainable change.

One of the tenets of RAMBO was to meet in personal spaces, rather than official ones—that is, in a team member's home, not at the university or in another work-related building. The team felt that being in a personal setting allowed people to be less formal and more open minded. They sought to create a warm and inviting environment that fostered friendship as much as science. As much as they loved RAMBO and the close relationships they developed over the years, there were challenges with being involved in the group. Many of the members were public employees at the federal, state, or local level. For daylong retreats, they had to take vacation time or sick leave to attend the meetings. Not understanding the value of this diverse group, some people were ostracized by their own organizations for participating in RAMBO. But, similarly to how time off for community service is now more frequently allowed in large organizations, being a member in a group such as RAMBO should be seen as valuable, and thus included as part of anyone's work roles and responsibilities.

The RAMBO group also learned and shared the idea that ecologically principled organizations and societies are the next best hope for the health and sustainability of humankind on our planet. They demonstrated that when professionals from many different disciplines—including ecologists, public health officials, and a wide array of other scientists—coalesce together, great visions arise. The group also understood that there is no formula or algorithm for creating a great organization, but if its members use all their senses, they know it's possible when they experience it. From their informal meetings, the group developed a set of RAMBO initiatives. The list included demonstration research projects on hantavirus and plague; a proposal to develop statewide infectious disease surveillance; a telecommunications network for New Mexico; assistance in developing a statewide immunization information system; and a proposal to test Telemed (a distributed database for clinical management systems) in the state's primary care settings. The group also sought to facilitate an investigation into an infectious disease cluster and an immunization strategy for the high-risk elderly in New Mexico.

Not to be kept to a single-page list of goals, the RAMBO group also sought to promote a Northern New Mexico Telemedicine Project; a statewide nosocomial (i.e., originating in a hospital) tuberculosis working group; a working group addressing the issues of privacy and the ownership of genetic and medical information; support for clinical database analyses by parallel data mining agents (PADMA); a Hispanic collaborative for research and education in science and technology initiatives; and a cooperative infectious disease project between Russia and the United States. These goals were formulated in the 1990s, during the first three years of RAMBO's existence, and many of them would still be considered progressive 25 years later.

The members lived by a quote from Dr. William J. Mayo, who stated in 1919, the last year of the Spanish flu pandemic, "Of all cooperative enterprises, public health is the most important and gives the greatest returns." Over the years, the RAMBO group continued to take on other important infectious disease health issues in New Mexico, including HIV, tuberculosis, plague, and hepatitis C. Anyone who has heard Simpson give a talk has heard him say, "Hi, I'm from New Mexico, land of the flea and home of plague."

One of the things that made this group extraordinary was their commitment to being special. To being extraordinary. They were driven not only by their love of science, but also by their desire to support each other. From the beginning, members of the core team were cemented in their belief that the other team members were worthy of their very best. That is, they were desirous of giving their best to the team, and everyone on the team was worthy of respect and kindness. Their priority was to create a collaborative learning environment that helped support each person with what they needed in order to do great work: vision, values, a growth environment, training, tools, time, and empowerment.

SYSTEMS THINKING

When I sat down with Gary Simpson and asked him about how they created not only a warm and caring group, but also a tremendously productive team for research and other initiatives, he started by describing the Navajo couple, Merrill and Florena, who were the first two hantavirus cases. He described the helplessness he and others felt at not knowing what was infecting these two young people, as well as not being able to stop the progress of the disease and the couple's subsequent deaths. Watching their lungs fill up with liquid so quickly, and not knowing what caused it, was heartbreaking. The duration of the outbreak was a stressful and painful time for the core team, but it was also transformative. Each member strengthened their feelings of compassion and empathy, and their ability to ask for help. Because the initial outbreak team and the RAMBO team could see the mechanics that had created the outbreak as they pieced the puzzle together, they became interested in learning about systems thinking and in using what they learned to create an extraordinary group.

In their research into systems thinking, they came across Peter Senge and his book *The Fifth Discipline*, which was first published in 1990. It described how organizations that are sustainably competitive are the ones that know how to learn, and such learning organizations understand how everything is interconnected. A change in one area will cause ripple effects in another area. Similar to the process that created hantavirus, systems are continually unfolding and in flux. The book's title points to five disciplines that are needed in a learning organization. The first is personal mastery; the second, a mental model; the third, a shared vision; the fourth, team learning; and the fifth, systems thinking. The RAMBO group digested the book and sought to create their own learning organization, based on Senge's ideas.

Commitment to the truth underlies the discipline of personal mastery. As Senge points out, commitment to the truth does not mean seeking the ultimate truth, but rather a relentless willingness to identify the ways in which we limit or deceive ourselves from seeing what is, and to challenge our theories about why things are the way they are—or appear to be. As scientists, the RAMBO team could get behind this idea and keep questioning their biases, searching internally and asking each other how systems create public health crises, and how the public health system could help reduce the occurrence of infectious diseases and other emergency health situations.

Mental models are conceptual frameworks, consisting of generalizations and assumptions through which we understand the world and then take action. Some of these mental models may contain limiting ideas or previously held wrong beliefs. Learning organizations facilitate learning at all levels and seek to question existing mental models. Back in 1993, I was a field technician helping to run two grids, each with 500 Sherman live traps for small mammals, on the plains of Colorado. Every morning I would go out to check the traps for rodents. We technicians would take multiple measurements on them and then let them go. At that time, my mental model was that I could not get an infectious disease from handling wild rodents and their droppings. Other than being bitten by one, or stepping on a rattlesnake, my perceptions of the safety concerns around wild rodents was low. I was wrong, as hantavirus clearly demonstrated. The team investigating the outbreak was constantly surprised by every new finding, but they learned from them. Each preconception was challenged and then changed.

With regard to the third discipline, the RAMBO team had a strong shared vision for reducing the threat of infectious diseases in the Southwest and promoting better health for everyone in New Mexico. The group also had a shared vision regarding the importance of science and the value of data. They all believed in supporting each other and being more than just colleagues. They were friends. And they understood and subscribed to one of Senge's principles: "That which is the most personal is the most universal." There was hardly any distinction between the team members' personal visions and their shared visions. They also understood that diversity strengthens shared vision, with these various aspects creating a single whole vision.

When I was in college in the 1980s, and later in the 1990s, one of my favorite activities was to walk into a classroom and see a screen and slide projector set up. It was even better if it was a science club or an informal meeting at someone's house, because it meant a slide show of a person's great adventure in the wild or their field season. I loved slide shows. I can remember the first time I went to Yates's house and saw the professional screen installed in his den, which descended with the press of a button. It was my first RAMBO meeting, and I remember thinking, "This is a man who takes learning to another level." The monthly RAMBO meetings centered around talks by invited speakers, presenting their research results in a fun and friendly atmosphere. It left an immediate impression on me, and I definitely wanted to be part of this group.

The group wholeheartedly embraced the fourth discipline—the idea of learning together—in this case by trying to solve public health problems. This mode of collective learning is sometimes referred to as a collective conversation.

Another component of collective team learning is understanding complexity, which is related to the fifth discipline, systems thinking. If you looked up the term "complex system," a classic example would be the ecology of zoonotic infectious diseases. It consists of one set of complex system on top of others. Climate, habitat change in response to climate, small-mammal and other wildlife responses to habitat change, and potential vectors (such as mosquitoes) adapting to a changing environment are all complex systems influencing each other. Then add human behavior and societies on top of this already complex system. This is an example of systems thinking.

The members of the RAMBO group sought to learn about systems thinking for learning organizations, in order to better understand the complex systems impacting human health. One of the core concepts of systems thinking is that there are different ways of looking at a problem. In a learning organization or team, a systems thinking approach fosters a change in mindset, going from seeing ourselves as separate from the world to being connected to the world and the overall system. We move from seeing problems as difficulties caused by someone or something "out there" to seeing how our own actions created the problems we experience. The RAMBO group wanted to become a place where people were continually asked to discover how they created their own reality and how they could change it, together. By being as diverse and multidisciplinary as possible, they were a group that understood what Senge meant with his law of systems thinking: dividing an elephant in half does not produce two small elephants. It's almost impossible for one person to see the entire elephant. The best approach is to have more pairs of eyes looking at the elephant from different angles and vantage points, communicating what they see to each other. Thus the most important lesson from the hantavirus outbreak was that solving this mystery required numerous disciplines to figure out what was killing people, how they were getting infected, why this outbreak was happening, and how to curtail it.

While the RAMBO group grew to be a learning organization, this facet was not the most important thing that made them extraordinary. What made them significantly different from other groups was the support and the caring for each other that they demonstrated. They understood that there are times when the best people, with the best intentions, making the best decisions possible at one moment in time, are going to have their big goals, or BHAGs, meet with failure. By supporting each other, they brought empathy, compassion, enthusiasm, and fun to each of their interactions. Due to the deep levels of expertise and experience of the group's members, they also added technical competence and accountability.

LONG-LASTING RESULTS

Few stories in life are a straight line leading to a happy ending. That includes this story. Life happens, and lives are changed dramatically. One of the things the RAMBO group wanted to understand was how to make their collaboration sustainable. But science teams, like most collaborations, have a finite existence. Projects end, people move on, and time passes. People move or retire or die. While the RAMBO group accomplished phenomenal things during the 25 years they actively met, most of that momentum has been lost. In particular, much of their spirit was abruptly taken away when Yates died from brain cancer in 2007. The memorial conference at his house that year was one of the last RAMBO meetings, although we did not know it at the time. The listserv for RAMBO still exists, and group members sometimes continue to share news about epidemiology and other relevant public health topics, but it is a shadow of its former self.

This story exemplifies the importance of leaders and champions for a cause. It takes a core person who will organize and continually bring people together, who keeps the team moving forward and communicating. Long-term collaborations need a leader who keeps asking, "What's next?" Passing the baton to younger researchers is critical in sustaining long-term science initiatives, but this is difficult, due to young researchers being forced to move around more in the worlds of academe and science. Most researchers now have multiple postdoc positions, and careers in science are harder than ever to navigate. Leaders in business and science have often wondered how to pass on their vision.

After Yates passed away, there were a few desultory RAMBO meetings, with an emphasis on getting a group together to work on geospatial epidemiology. One of the memories that Gary Simpson has is of a researcher coming to RAMBO to present a talk on the High-Performance Computing Center at the University of New Mexico. After the formal presentation, the speaker mentioned the vulnerability of their computers to viruses and cyber attacks. One of the RAMBO disease experts responded by offering that it was "too bad you don't have an archive in the High-Performance Computing Center that acts as its own immune system, because if future attacks come that are similar or related, like the immune system, this archive could develop a repertoire or full coverage of immune responses based on challenges." Six months later, the computer researcher had taken that comment to heart and had worked out the philosophical construct of an immune response to create modules for the High-Performance Computing Center, which was subsequently funded by a grant. The RAMBO group knew that the connections and synchronicities they had were something magical, but it was not sustainable into the next generation. Soon its members stopped meeting regularly.

One of the primary lessons learned from the 25 years when RAMBO was operating is that relationship-building is critical for scientific collaboration. The combination of connection, trust, and friendship is the secret ingredient for transformational science, and being in the Yatesian Zone is where the synthesis and integration of ideas that can move science forward occur. This not only increases the probability for scientific achievement, but also for the creation of more happiness in life. Combining vision and happiness is where magical things happen. The study of the benefits of connection with others is a research field in itself. Connecting with others requires being open, vulnerable, and available to another person. Being open to new ideas and thoughts is also essential for transformative science. The key ingredients of human connection are empathy and compassion—a sense of goodwill. The basis of the scientific method is building on and often refuting or disproving ideas, in order to keep focusing on the framework of our understanding of the world. How we do that together, and how we treat each other well, can move us further and faster forward into a brighter future. The good news is there are many strategies and ideas available that can be applied to create better team collaborations. The next chapters highlight some examples of both good and bad collaboration and the best practices available, starting at the highest community-level scale and continuing down to the intimacy of two people on a team.

CHAPTER 2
COMMUNITIES

The greatness of a community is most accurately measured by the compassionate actions of its members.

CORETTA SCOTT KING

I have often wondered if polar bears reminisce about the good ol' days when there was lots of sea ice. Or do they wake up each morning and try to go about the task of finding their next meal, not noticing how things have changed? We know that over the last 36 million years, much of the Arctic has been covered with ice, with only some intermittent gaps. One of those gaps occurred about 5.3 million years ago, when there were dense pine forests in the northern Arctic. At that time, Greenland had 30% less ice than today, and global seas were about 60 feet higher. Average temperatures back then were about 4° Celsius higher than our average temperatures today, and that much of an increase is expected again by the end of the twenty-first century. It is estimated that about 4 million years ago, Arctic Sea ice and temperatures were similar to what they are today, and summers were probably ice-free at the North Pole.

Scientists use orbiting satellites to track the extent of sea ice in the Arctic year round. The minimum amount of ice in September has continued to decrease over the past four decades. Some projections suggest that by 2050, the Arctic will be ice-free during summers. In 2013 the North Pole was unfrozen, with open water. While this has occurred at other times in the past, its frequency is leading scientists, governments, and industry to start planning on being able to navigate across the Arctic on a regular basis.

But we are not here to talk only about polar bears. We are here to discuss the top challenges that the world faces in the twenty-first century, and to examine how scientists and the research they do can help address the threats that may arise in our collective future. We will look at how scientists can work better together by building trust, increasing empathy, and improving communication across the board—in other words, have kick-ass dream teams. While we can advance the technologies and increase the sophistication of the analyses researchers use, we can also make great leaps forward in how we collaborate in science. By making improvements in the areas of relationship-building, trust, communication, and compassion, we not only can have careers in science that are more fun and rewarding, but we can also increase creativity in problem solving, find solutions fasters, augment scientific productivity, and attack larger challenges that need a multidisciplinary effort.

About 2.6 million years ago, a geologic uplift narrowed the Arctic Ocean gateways, such as the Bering Strait, which isolated the Arctic region. The Arctic Ocean's circulation was restricted, causing a buildup of fresh water and creating conditions favorable for major ice sheets to form. At that point, runaway cooling began, with ice sheets extending as far south as present-day St. Louis. That was the most recent cycle of ice ages, and our human ancestors were forced to adapt. Around 200,000 years ago, what we call "modern humans" emerged. Then, approximately 12,000 years ago, this ice age ended, which helped modern humans develop civilizations.

In 1958, scientist Charles Keeling started measuring atmospheric carbon dioxide (CO_2) at the pristine Mauna Loa mountaintop observatory in Hawai'i, and the world has watched the gradual increase in atmospheric CO_2 every year since. That first measurement of the CO_2 level was 316 parts per million (ppm), a little higher than the preindustrial level of 280 ppm that appeared to be stable for the last 6,000 years. In 2018, atmospheric carbon dioxide reached 400 ppm for the first time in at least a million years. This curve became known as the Keeling Curve, and it is the second-most recognizable representation of climate change, next to polar bears.

Keeling noticed an obvious pattern: CO_2 levels are highest in the spring, when decomposing plant matter releases it into the air, and they are lowest in autumn, when plants stop taking in CO_2 for photosynthesis. He could see that nighttime air contained a higher concentration of CO_2 compared with samples taken during the day. He realized that plants take in CO_2 during the day to photosynthesize—or make food for themselves—but at night they release CO_2. Climate science has grown in the last six decades to become a sophisticated and complex field, filled with terabytes of data used in forecasting models. Thousands of researchers around the world contribute to identifying the patterns and documenting their impacts. These impacts can be as tiny as a tanager arriving a few days earlier in the spring at its breeding location in the north, or as dramatic as the loss of sea ice.

Arctic temperatures have warmed at double the pace of the rest of the planet, leading to the expanse of frozen seawater off the Arctic Ocean and neighboring seas shrinking and thinning in the past three decades. Recently, NASA scientists found that since 1958, the sea ice cover in the Arctic has lost around two-thirds of its thickness. Now, 70% of the sea ice cap is seasonal ice, or ice that forms and melts within a single year. At the same time, the sea ice is also thickening at a faster rate during winter, even though it is vanishing quicker than has ever been observed before. This may seem counterintuitive. How does a weakening and disappearing ice cover manage to grow at a faster rate during the winter than it did when the Arctic was colder and the ice was thicker and stronger? As my former Comparative Tropical Plant Ecophysiology professor answered to most questions asked in the class, "It is a complex phenomenon." Like most systems in nature, biology, business, and human cultures, there are often negative feedback loops. Global climate models seem to do a good job of capturing the status of the Arctic sea ice, and they point to a bigger player in the sea ice–thickening game. Ice is an insulator that keeps energy from passing back and forth between the atmosphere and the ocean, and the thicker that ice is, the better it insulates. Thinner ice melts and regrows faster than thicker ice, and it's also weaker, so when the wind blows, the ice piles up into thick ridges. The Arctic ice pack is a much more dynamic environment now than it was when it was thicker.

In 2012, an extensive and thick area of sea ice off the Alaskan coast had delayed the final yearly fuel delivery to Nome, Alaska, and a US Coast Guard icebreaker needed to create a path through the ice for a Russian oil tanker. For situations like this—a special mission and extra-thick sea ice—highly accurate predictions of sea ice change are critical for the safe passage of ships. The National Ice Center, which is cosponsored by the US Navy, the National Oceanic and Atmospheric Administration (NOAA), and the US Coast Guard, produces analyses forecasting sea ice–covered waters. For this sea ice prediction mission, the National Ice Center team uses satellite imagery that feeds into CICE, also known as the Community Ice Code, also referred to as the Los Alamos Sea Ice Model. This acronym is pronounced "sice" (not "c-ice"), to avoid confusion.

The computer code for CICE was developed through years of community collaborations that solved a collection of mathematical equations representing physical processes that occur during sea ice evolution: the growth, melting, and movement of the ice, along with the snow and meltwater carried with it. While we (and polar bears) care a lot about where sea ice *is*, it is more important to understand where this ice *is not* by modeling the physical system of sea ice. The open water between ice floes controls the flow of moisture, energy, and momentum between the atmosphere and ocean, as well as the snow and ponds of water on top of the ice. The major components of the CICE model are thermodynamics, momentum, and horizontal transport routines between the snow and the ice.

CICE was first released in 1998. It was developed at the US Department of Energy's Los Alamos National Laboratory by Dr. Elizabeth Hunke and her colleagues. The first CICE community user was the Naval Postgraduate School. Soon other national and international scientists began employing and then contributing to the code. Today, CICE is a sea ice model used in more than 20 countries for applications as diverse as global climate projections for the Intergovernmental Panel on Climate Change reports; research into fundamental sea ice processes; daily sea ice and weather forecasting; and special missions, such as fuel deliveries to Alaska. The CICE Consortium is supported through the US Department of Energy's Office of Science; US Department of Defense; National Science Foundation; National Oceanic and Atmospheric Administration; and Environment and Climate Change Canada. In 2021, the CICE Consortium received one of the prestigious R&D 100 Awards for the scientific achievements of this community of sea ice researchers.

Hunke works in a nondescript 1950s office building with painted cinderblock walls in the middle of Los Alamos's primary technical area. There are many hallways, offices, and breezeways to other buildings, plus a hidden hallway to the second floor. The first time I met Hunke in her office I was a few minutes late, wandering through the building in circles and finally stopping to look at one of the emergency maps posted throughout the structure. Visiting Cold War–era government science buildings is like stepping back in time. It is easy to forget the history that might have taken place in those nondescript rooms.

There are two types of research careers. One is where individual scientists adapt and evolve their research over time to meet new pressing needs or basic science questions, as well as adjust to the flow of funding. The other is where a scientist is committed to solving a central question, adding more to that specific foundation each year. Hunke follows the latter path, spending the majority of her career modeling sea ice.

Hunke wanted to apply mathematics to real world problems. Her original interest was in understanding hurricanes. As she put it to me, "Hurricanes are pretty cool." As a pandemic researcher myself, I understand what she is saying: the underlying mechanisms and dynamics of some pretty horrible challenges to humans and the planet can be fascinating. As a graduate student at the University of Arizona, Hunke began applying numerical models to understand hurricanes, and soon an opportunity came up to work on the frozen part of the ocean, sea ice, as a postdoc at Los Alamos National Laboratory. This community of researchers wanted to understand the variability of the sea ice system—ranging from wind pressures and how cracks in sea ice form to how the model for sea ice interacts with changes in the oceanic and atmospheric models. While writing the computer code for these extensive sea ice models kept Hunke interested in solving the complex phenomenon of sea ice, it was the community of other scientists around the world that kept her invigorated.

THE NEED FOR COMMUNITY

Researchers agree that humans are a cooperative species, and community gives us a sense of belonging. People choose to work together not only for selfish reasons, but also—as Samuel Bowles and Herbert Gintis pointed out in *A Cooperative Species: Human Reciprocity and Its Evolution*—because we are "genuinely concerned about the well-being of others, try to uphold social norms, and value behaving ethically." They argued that we developed these moral sentiments because our ancestors lived in environments where groups of individuals who cooperated tended to survive better and reproduce. The challenge of the day 70,000 to 90,000 years ago was to find food and shelter, escape accidental death, and survive attacks by animal predators and rival groups. The challenges of today are at the planet level, complex enough that individual minds cannot fully comprehend them. To solve the predicaments that the planet now faces, we must have stronger collaborative bonds and communities.

A community is an organization or group where there is effective and transparent communication, equality, respect for differences, high levels of cooperation, and a significant degree of trust. When research project leaders make a concerted effort to create a sense of community among their team's members, productivity and efficiency go up. Communities, even in science, are not without conflict, but their members have enough dedication to resolve their differences and use them as learning opportunities. Conflicts in research often lead to transformational science. Yet it is the "how" of dealing with conflicts among team members that is important. The focus of a community should be on a vision of the future that can be created together, and the actions needed today to get to that future.

COMMUNITY BUILDING

A scientific community is a diverse network of interacting researchers working in a particular field. Its purpose can be defined by the situation it confronts. But the sense of community within a team is the feeling that people have of belonging to a group and being important to one another. That sense is the magic that can allow ordinary teams to do extraordinary things.

CHARACTERISTICS OF A GOOD COMMUNITY

Social scientists have been researching the attributes of communities for decades. In describing the determinants of what makes a good community, two studies are particularly relevant. In 1986, David McMillan and David Chavis published an article entitled "Sense of community: A definition and theory." They found that four factors show up as attributes of what we would call a "good community." First, there is membership—the feeling that we are invested in the community, and that we have a right to belong and feel welcome. Second, there is influence—the sense that we have some input in community issues that affect us, and that our perspectives are respected and appreciated. Third, there is integration and the fulfillment of needs. This attribute is based on the notion that the community offers opportunities for both individual and social fulfillment, including basic needs and social interactions. Fourth, there is shared emotional connection. This is based on the shared history or sense of the community, and the quality of interactions within it.

The second study comes from the Soul of the Community Project, conducted in 26 communities across the United States by the Knight Foundation and Gallup in 2010. The aim of this research was to look at those factors that facilitate "community attachment." Besides highlighting individual elements, they found that those communities with the greatest levels of community attachment also had the highest rates of growth in their local gross domestic product. The community characteristics that most influenced community attachment (in order of importance, from highest to lowest) were social offerings, openness, aesthetics, education, and basic services. While there were some differences in the relative strength of each of these factors across the 26 communities, all five had the most influence on feelings of attachment within a community. Another five important, but somewhat less influential factors were leadership, economy, safety, social capital, and civic involvement.

At this point you may consider that the second study was about the larger communities that we live in, which do not form a science team. The sense of community, however, is what we are interested in building in both our smaller teams and in larger science communities, such as the sea ice modeling community. We can see that social connections, transparency within the group, continual learning, and how the project supports their job or research work are key aspects for people, allowing them to feel engaged with the team. We can also see that leadership is an essential factor, and I would argue that for smaller teams, the importance of leadership should be highest on the list. Feeling safe within the team, having adequate funding, enjoying the social capital of collaborating on projects with high visibility or stature, and working on research questions that have meaning and are striving to solve challenges the world faces are also important for creating a sense of community in science teams.

The community of sea ice modelers, which later became the CICE Consortium that emerged in the mid-2010s, has many of the community attributes noted above. The CICE Consortium's vision is to enhance the development of sea ice models for and by the modeling community. The CICE Consortium has described three goals: the acceleration of scientific development, the acceleration of R&D transfers to operational use, and the development of a vehicle for collaboration and sharing. Community development is the core process for their improvements in the modeling code, confidence testing, and scientific oversight. The CICE Consortium team maintains its vitality and cohesiveness through monthly teleconferences (with other meetings scheduled as needed), and through communication via tools in GitHub (an internet hosting service for software developers). Team members also strive to maintain the consortium's visibility in and interactions with the larger scientific community.

LEVELS OF COMMUNITY

The smallest scale for a research collaboration would be two people working on a project together, and the largest would be an interagency, international community of stakeholders. In the case of the CICE Consortium, we are talking about a scientific collaboration at the biggest scale, with layers of communities (figure 2). But the CICE Consortium also exists at a smaller scale, in the layers of communities concerned with modeling sea ice. The consortium's mission is to foster collaboration on sea ice model developments for Earth-systems research and operational applications.

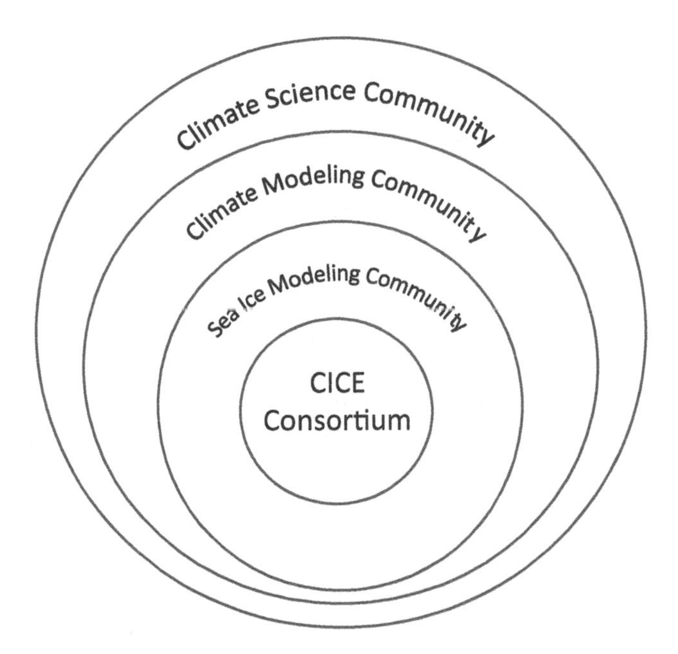

FIGURE 2. This model is useful for understanding the layers of communities that make up the larger scientific community that models sea ice and works in climate change research around the world. *Courtesy of the author*

As one goes up in scale for a scientific collaboration, the challenges increase. Those for multi-agency working groups of scientists and other stakeholders are extensive, and they reflect the complexities involved when scientists working for different institutions or agencies engage in collaborative ventures. In a study done in 2002 for the National Foundation for Education Research in the United Kingdom, the authors found that the main challenges of multiple agencies working together were centered around the areas of funding and resources, roles and responsibilities, competing priorities, communication, professional and agency cultures, and management. For the hantavirus team and RAMBO, which did not receive the blessing of some of the institutions involved to have their members collaborate, this could be considered more of a roadblock than a challenge.

For collaborations at this highest scale, the dissemination of information and the need to keep everyone in the project informed of developments is critical for the success and sustainability of a community. The 2002 UK study on multi-agency working groups also found that communication within these larger communities operates at three broad levels: procedures, systems, and building a sense of community. Procedures include developing a language that establishes roles and responsibilities, and setting clear communication frameworks that cut across an agency's separate "swim lanes." Procedures also involve interagency communications, such as through a newsletter or website, that keep the researchers updated. Another example would be to write clear protocols and principles for communication within the community, so researchers are aware of their responsibilities to others.

At the systems level, setting up specific information-transfer modes makes it easier for communication to take place. In the "new normal" of working remotely, there are vast quantities of software programs and cloud-based computing sites to facilitate collaboration. These are critical for science teams and related communities, and researchers will have their favorite collaboration platforms. There are other systems, however, that can be used to improve communication, such as setting up an internal email list for the group. The key is to continue to upgrade and check the level and effectiveness of communications within the community.

The third level of communication is building a sense of community at the human-to-human level. This includes talking to others frequently (face-to-face, if possible) and socially. If the community, like the CICE modeling community, only meets annually at a scientific conference, then making time for social gatherings is important. It is over dinners and drinks after the day's formal sessions are concluded that the most valuable relationships are forged. While teleconferences, or telecons, were available as far back as 20 years ago, in the age of the COVID-19 pandemic, we have taken online conferencing to a new level. We may be 4,000 miles apart, but video communication when working at home has led to a new level of intimacy that we did not share before. We now may know that a colleague has a 4-year-old who likes umbrellas or a cat that is hungry all the time, or that they play an instrument. And we all try and read the titles of the books on shelves located behind the person we are speaking to.

One of the core values of the RAMBO group was to focus on taking the time to listen to the perspectives of others, especially researchers working in other fields or agencies, knowing that they might bring new insights into an old problem. Diversity and the inclusion of all types of members helps establish a forum for self-reflection, allowing us to question our own technical assumptions, or even any stereotypical biases we may have had toward other scientists or fields.

Another strategy for improving communication within a community is to identify key people in each agency or organization who can communicate well across agencies, at both strategic and operational levels. Continual work on the CICE model has improved our ability to understand a type of anchored ice, called "land-fast ice," attached to the shore or sea bottom, which can block shipping lanes and northern ports. As Hunke points out, the polar regions are not desolate. Instead, they are bustling with shipping, energy development, fishing, hunting, research, and military defense operations. Sea ice makes navigation hazardous, as thick ice can block boats and make it difficult for US Navy submarines to surface during an emergency. The CICE model is essential in helping agencies predict polar sea conditions, which they need to understand in order to develop infrastructure, shipping, and other transportation plans. This new tool can also provide insights into the complex ecosystems within the polar region, such as the algae that are essential for some polar creatures.

A CICE submodule, known as Icepack, has also been developed by Hunke and her team. Icepack contains the single-column physics and biogeochemical aspects of CICE, such as ice ridging, thermodynamics, and hydrology. With these features now independent of some other types of dynamics, they have become easier for a greater number of modeling groups to use, and both CICE and Icepack have demonstrated an ability to accurately simulate the characteristics of sea ice.

The CICE model was developed and maintained by the US Department of Energy as an efficient sea ice component for use in coupled atmosphere/ice/ocean/land global circulation models. Over the past two decades, a broad community of climate and weather forecasting groups have adopted and enhanced the code. The consortium has become a vehicle for collaboration for sea ice–model support and development as the larger community continues to use and improve the model. The consortium is set up as a framework of people able to evolve general code for sea ice models. To be a part of the inner CICE Consortium, member organizations must contribute approximately 25% of a full-time equivalent position.

Many layers of science communities surround the CICE Consortium. A community of scientists and stakeholders model sea ice in the polar regions and may use the CICE model. A community of global climate modelers construct models of the oceans, the atmosphere, and other Earth systems that may link to or depend on the sea ice models and help in our understanding of the annual changes to the Earth's sea ice. The community of other climate researchers include experimentalists and field scientists who are collecting data through instrumentation or are out in the field in some of the world's harshest conditions. Each of these communities is connected through either formal or informal networks.

WAYS OF BELONGING TO A COMMUNITY

Charles Vogl has been working to build communities since his time as a Peace Corps volunteer in Zaire. He has written two books about building and cultivating strong communities. In *The Art of the Community*, when looking at ways to strengthen communities, he laid out seven principles for developing a sense of belonging in communities. The first principle is boundary, or the line between members and outsiders. The second is initiation, or the activities that indicate a new member. The third is ritual, or things we do that have meaning. The fourth is the temple, or the place set aside where we find our community. The fifth is the stories that we share, which allow others and ourselves to know our values. The sixth is our symbols, or the collection of things that represent ideas that are important to us. The seventh is inner rings within our communities, or our path to growth as we take part in the group.

If we look at the CICE Consortium and its sea ice modeling connections as strong communities in their current state, we can see how well they adhere to these seven principles. In my conversation with Charles Vogl about his ideas on communities, however, he told me that "there is no reason to have all the principles in a community," which would be "like ordering a dinner at a restaurant and wanting something baked, fried, roasted, fresh, sautéed, and frozen all at one meal." What the seven principles can do is give us a framework for thinking about the *how* of creating a strong community, or of building a sense of community in a team.

With regard to the boundary principle in a science community or team project, there is usually a clear line demarcating whether a person is part of the community. For the CICE Consortium, this is either a 25% full-time equivalent position or an equal amount of in-kind support. For the sea ice modeling community, it means being an active user of the sea ice model or its results, as well as participating in the community's online forum or in sessions at a science conference. For scientific societies, members are those who pay annual dues. For research projects and collaborations, membership can be less clear, as there may be participants who are left out of group communications, or specific roles and responsibilities might not be delineated. People can be either consciously or unconsciously left out of research teams, even though they may actually be contributing to the group. In some cases, I have seen colleagues "ghosted" from research teams. They were contributing during one phase of a project, and then, with no explanation given, were not included in the rest of the project or left out of publications arising from it. In these situations, people tend to move on, but the impacts can be hurtful and long lasting. It is not a good way to manage the boundaries of who is on a team. Effective strategies for having "crucial conversations" within a team exist, and this book will go over them in a later chapter.

Initiation refers to activities that apply to a new member in a community. For larger scientific communities, like the sea ice modeling one, initiation is the feeling of being a part of the community. This happens at conferences or meetings of the community, and its extent varies, depending on how inclusive and welcoming the community may be. Some societies have now taken to writing diversity and inclusion statements. For example, in 2015 the American Ornithological Society published a diversity statement that it uses to promote the code of conduct of the society, as well as for its annual conferences. It is with permission that I share it here.

> The greatest asset of the American Ornithological Society (AOS) is the diversity of individuals representing the regions where they work, the disciplines that comprise their research, their individual viewpoints, and their generosity of knowledge and time in advancing a global perspective in ornithology. The mission of AOS is best fulfilled when we embrace diversity as a value and a practice. We maintain that achieving diversity requires an enduring commitment to inclusion that must find full expression in the culture, values, norms and behaviors of AOS. Throughout AOS's programs, events, publishing, and professional development activities, we will support diversity in the membership, leadership, volunteers and employees in all its forms, encompassing but not limited to age, disability status, economic circumstance, ethnicity, gender, race, religion and sexual orientation. Leading by example, AOS aspires to make diversity a core and abiding strength among our membership.

Rituals are the things we do that have meaning and may identify particular events. They have a prescribed order, so people can depend on them or use them to celebrate certain activities in life. We tend of think of science as not being ritualistic, but the hooding ceremony for a freshly anointed PhD or the continuity provided by a Thursday afternoon seminar series would tell us otherwise. One of the most important aspects of rituals is that they not only mark time, they *create* time. By defining the beginnings and endings of developmental or social phases, rituals structure our social worlds, including how we understand time, relationships, and change. Most research projects will have kickoff meetings, but far fewer have gatherings to signify the end of a project. For communities like the sea ice modeling one, getting together in a session or a working meeting at the annual scientific conference and then going out to dinner and, later, to a bar, constitutes a key ritual. These activities offer an opportunity to keep on building relationships within the community. There is also the ritual of weekly team meetings, or quarterly meetings if it is a larger community. Sharing safety concerns can now be seen as a ritual to start the meeting. During the pandemic that began in 2020, the world learned even more about the importance of rituals in our lives. Performing a PhD hooding via a Zoom meeting and moving our scientific conferences online kept us grounded. Not only was the weekly lab meeting telecon still important, but so were the Friday afternoon clubs, where we met in person and felt connected to our fellow colleagues.

The temple is the place where we can find our community. Charles Vogl borrowed the name from religion, where it connotes a special meeting place. This may be in the laboratory or conference room where the team gets together. Or it could be the GitHub site, like the one the sea ice modelers have, where the community meets to discuss new code and analyze ideas. As scientific teams have become larger, and their collaborators more spread out, the temple has moved to online collaboration sites, such as Slack, Confluence, Microsoft Teams, and dozens of others. Having a place to share ideas and information is critical for research teams and communities. For some communities that meet annually at conferences, the most important temple will be the hotel bar, where conversations can extend into the wee morning hours.

Stories are what we share that allow others, and ourselves, to know our values. They are what are talked about in the hotel bar. They connect us and make us laugh. At my own ornithological conferences, there are now storytelling events that offer the attendees an opportunity to relate their personal tales. Stories can recount lessons learned about how not to combine two chemicals in the laboratory and thus avoid an explosion. They can describe the struggle of being a scientist in a competitive world that has limited resources and sometimes includes challenging personalities. Stories build trust and create common ground for the things in life we worry about and grapple with as scientists and human beings. CICE community members share their code development strategies and results on their GitHub site, but they save their more intimate stories about the struggle of working on climate change and the challenges of different political ideologies regarding climate for when they get together in person.

Symbols represent ideas that are important to us. In business, a logo is a critical component in recognizing a company or in branding a business. Even NASA has a special patch, with emblems designed and worn by astronauts and other people affiliated with a space mission. It is now common for groups to have logos for their project, or stickers that mark equipment and/or indicate being part of a community. Symbols represent what is important to us, although they may change over time as our priorities shift. Thinking about which symbols represent a community or team can help create a sense of identity and highlight the group's values. In science, there is a plethora of potential symbols to choose from (e.g., chemical symbols), and we often use symbols within symbols, indicating our nerdiness.

Inner rings represent the paths to growth we take in science as we participate in our particular group. While the inner rings of a community may appear to have a hierarchy, they generally do not. These rings are based on experience, expertise, skills, and wisdom. As we progress through life, we enter into the inner rings of various communities. In some ways the rings indicate tribes within communities, based on these growth factors. Strong and emotionally connected communities have different layers of inner rings that allow members to enter, gain more experience, and grow. A good example is allowing students to be a part of a faculty leaders' meeting. Or having diverse representatives from all levels serve on committees. This inclusion has many benefits. It allows new growth, provides leadership experience, and encourages respect among those in different rings. Vogl pointed out that it is not important for each member of a community to pursue these inner rings. A strong community offers a panoply of opportunities for growth and new experiences, and it allows all people within it to develop relationships with each other.

Most scientific societies have a mission to include students, as do science projects. There is a culture in research for educating the next generation of investigators, even those outside of traditional universities. Aside from the fact that students have the time and, often, the best grasp of the newest technologies and computer skills, they are relegated to much of the day-to-day work required in research. For larger communities in science, seeing leadership and community organization modeled by its senior members is an essential part of growth. Organizing a conference for a scientific society, or being a member of one of its committees, are valuable ways for younger researchers to interact and work with more senior members in the community, outside of their local laboratory environment. Seeing how such leaders come together offers examples of people forging ahead through challenges and finding solutions. If the community is thoughtful about this, then the behaviors that are modeled should be ones of respect for others, honest but kind communication, and the inclusion of every idea. Not all cultures in science are based on these behaviors, but there is strong evidence that this is changing as science evolves, with the next generation becoming more conscious of how we interact and collaborate.

There is also less tolerance of bad behavior within the communities in science. A good example of this is that many of these communities and science societies now have a code of conduct statement or an ethics committee. The latter is responsible for developing a process where society members can come forward with issues of concern or grievances about other members. This quickly can become sticky ground, but having a means of investigation is important. It allows issues that would previously have been swept under the rug to be dealt with and resolved. One characteristic of a strong community is having a way within it for all the inner rings to talk about and resolve issues. This builds trust among the rings. The most important part of this trust component is that the process will be carried through to completion, with at least some communication of the resolution being given to and acted upon by the impacted parties. Depending on the situation and various legalities, the issue may then be used by the whole community as a lesson learned.

WHAT COMMUNITIES CAN ACCOMPLISH

As science evolves and communities mature, the ways in which we "do" science grow and get better. Communities are evolving to have more integrity. As Dr. Henry Cloud pointed out in his book *Integrity*, the Hebrew word for "virtue" is close in meaning to "integrity." Resilient communities are communities that are built on trust among the inner rings and have given thought to defining the integrity and the values of the group. Communities that have trust and communication among members in their inner rings will then create greater trust among all of that community's stakeholders—and, thus, more overall resilience and sustainability.

Communities can ask questions that increase the connectedness of their inner rings and the sustainability of the group. What are the layers of leadership and seniority within a community? Does it have a sustainability plan? How will the community respond to a major disruption, such as a pandemic? Are new leaders and champions being developed to take over the baton? What growth opportunities are being offered to all members of the group? For some communities there are hard endings on the horizon, and for others this may not occur. In the sea ice modeling community, the models they use will become obsolete. Their computer code will become antiquated, with better codes and forecasting tools becoming available. Like wildlife and plants that adapt (or do not) to climate change, communities will need to adjust and evolve as scientific fields change. Sustainable communities are those that can do so, as well as keep working toward inclusivity and diversity.

John Maxwell is a top authority and prolific author on leadership. He often has talked about the three things that every person asks themselves when they meet a new person or ponders throughout a relationship, either consciously or unconsciously. Do you care about me? How will this benefit me? Can I trust you? Being part of a community is entering into a relationship with it. When we think at the community level, we must ask ourselves the same questions. Do we care about our members? How does the community benefit members inside and outside the group? Can the members of the group trust the larger community?

Developing authentic communities that serve organizational visions and goals is worth the investment in time and other resources. Communities create new values through innovations, which help retain stakeholders in the group. Communities also help in the recruitment of new scientists and create fundamental shifts in scientific fields, all of which keeps them moving forward. In the past, a group coming together in agreement or disagreement on a topic has made all the difference in settling debates in science, such as when leading scientific organizations, like the American Association for the Advancement of Science (AAAS), have posted position statements on, for instance, the human causes of climate change. Communities allow back-and-forth communications among members, enabling them to look at all sides of an issue.

In the past three decades, technological advancements in computers, analyses, and instantaneous communication around the globe have led to fundamental changes in science. With the hyperspeed of communications and interconnectivity in the world, it is faster and easier to have scientific results become available to all. This scientific revolution has made it even more important for us to focus on how we "do" science. Developing trust and integrity has also become more and more vital. Science has changed in both predictable and surprising ways. This has had remarkable effects on how we collaborate, ranging from the larger community level, such as the sea ice modeling community, to the smallest of collaborations between two people.

Exercise

Here are some questions to ask yourself and to think about. You might want to try writing down your thoughts in a journal.

1. What types of communities in your life do you belong to, and why?
2. Do you prefer to work with smaller or larger communities? Or to work alone?
3. Can you think of an example of a challenge in science where a larger community is required? And one where it is better to have a small team?

CHAPTER 3

A SCIENTIFIC REVOLUTION

> When you see something fun and interesting, just go for it. Dive in and do it.
>
> ELIZABETH HUNKE, *sea ice modeler*

In the last 20 years, the advent of new technologies, increased computational power, and big data have not only changed science, but they have also altered scientific collaborations. How we do science has expanded in the past two decades compared with the last 2,000 years of its history. While the scientific method has stayed intact, how we collect, store, and analyze our data has changed. New technologies have allowed us to see deeper into DNA and make changes to the genome. Computer processing speeds and storage space have followed Moore's law, with the number of transistors in a dense integrated circuit doubling about every two years. Analyses and modeling in research have become faster and more complex every year. Statistical and specialty analytics, like bioinformatics or survival analysis programs, are more user friendly. New scientific journals pop up each week, although many of these are predatory and do not adhere to a critical aspect of publishing: peer review. With these advances in science, the expectations of researchers have also increased, ranging from the number and quality of manuscripts published each year to the total amount of funding obtained. Scientists no longer are expected to have projects; they are expected to build "programs." The goal of this book is to present strategies and ideas to help unleash the true potential for research collaborations and increase the effectiveness of all types of science around the world.

HOW HAVE RESEARCH TEAMS CHANGED?

The rapid and recent changes in science have also led to adaptations in how we collaborate with each other and work most effectively as science teams. In the past decade, such questions have led a suite of researchers to start investigating the science of team science. Researchers and authors in this new field now point out that scientific collaboration has gone through a revolution. In their book *The Strength in Numbers: The New Science of Team Science*, two of the founding thinkers and investigators for understanding how researchers collaborate—Dr. Barry Bozeman, a professor at Arizona State University, and Dr. Jan Youtie, a professor in the School for Public Policy at Georgia Tech—defined such collaborations as "the social process of bringing together human capital and institutions in the production of knowledge." They also pointed out distinct characteristics of the recent revolutionary changes in scientific collaborations. First, the total number of collaborations has expanded, as has the number of team members per collaboration. Second, gender diversity and multiculturalism have increased on team projects. Third, the number of global and international research teams has grown in the last 20 years. Fourth, not only are there more multidisciplinary research teams, but there also is an active call to fund multidisciplinary work that addresses larger and more complex scientific questions. Fifth, ethical issues and coauthorship concerns are given more attention and press. Sixth, with the recent rapid growth in the commercialization of science in academia and industry, there has also been more thought about team science, with new policies for managing collaborations being put in place.

COLLABORATION SIZE

Science teams appear to be much larger than they were 20 years ago, and there is now a greater expectation for research to address larger questions with the same amount of funding—or sometimes less. The need to comply with institutional research boards and similar committees, as well as stricter financial requirements, has also added complexity over the last two decades. The motto "It takes a village" can also apply to research projects. The community necessary for the success of a scientific project requires more people and includes administrative professionals, safety and compliance specialists, communications personnel, and even the janitors who clean our laboratory spaces. While we may complain about the bureaucracy and the burden of what it now takes to be able to do research safely and securely, we also need to remember that the people whose job is to support science are also part of our team. As Keith Ferrazzi and Noel Weyrich pointed out in their classic book, *Leading Without Authority*, the old workplace rule limited a team to those who reported to a leader, while the new rule is that your team is made up of everyone—both inside and outside the institution—who is important in achieving the goal of the project or mission. Many of the points discussed in the present book will follow this new rule, which is critical for a project's success, and be applicable to fruitful collaboration not only within science teams, but also beyond a team's boundaries.

The still-standing record for the number of authors on a paper is 5,514, in a 2015 article by Aad et al. (the entire group was referred to as the ATLAS and CMS Collaboration). These individuals were from the teams at the Large Hadron Collider who collaborated in finding a more precise estimate of the size of the Higgs boson. Only the first 9 pages of the 33-page article described the research and listed the references. The other 24 pages contained the coauthors and their institutions. While this extreme example represents an inclusive approach to authorship, it also represents different teams of people working together toward a single research mission. Even ignoring this outlier example, we have seen an increase in the size of research teams, represented by a growing number of authors in the papers resulting from their work.

Why has the size of research collaboration teams expanded so much over the last 25 years? Is it the increasing ease of communication through the internet and other communication technologies? Is it that the questions scientists are tackling today are more challenging and need multidisciplinary approaches? Is larger team size driven by the increasing preponderance of research in the health and medical fields? The short answer is yes, all these things have led to larger collaborations. While we can agree that there is a positive benefit to bigger and more multidisciplinary collaborations, there are also costs associated with larger teams. The time and effort involved in proposing more complex and difficult projects, and the amount of coordination required to set up and manage larger teams, can occupy more time than it takes to work on the actual science. There are also risks in designing more difficult questions that need larger teams. Larger projects have greater costs, and it can become harder to find research agencies or sponsors able to fund such projects, while competition for these monies increases.

While the size of collaborative groups has grown across most, if not all, disciplines, some fields have not expanded as much as others. Much of this is driven by funding amounts that have remained low in some fields, while increasing exponentially in others, such as health and medicine. For smaller, specialized disciplines within a broader field (like my own, ornithology), funding often comes from more specialized granting organizations or clubs. Sometimes, it amounts to just a few thousand dollars, only enough to cover the cost of a graduate student's gas to get to a field site. For a team with only two researchers, the field assistant is often a volunteer hoping to gain experience. In graduate school, most projects are designed to be done alone—going from the project idea, to data collection and analysis, to the final manuscript. Yet with the revolution in science collaborations taking place, many graduate students are having to learn the ins and outs of collaboration while also learning how to do science.

Regardless of the reasons why collaboration teams have increased in size over time, the goal of this book is to offer suggestions on how to work more effectively and painlessly with others on projects. The more challenging and complex the issues we face in the world are, the greater our need for larger teams to tackle these questions. Through such increasingly collaborative groups, we can learn to have more efficient, productive, and meaningful projects that could make the world a better place. When we learn to build trust and to communicate better with our colleagues, then we can have more fun working together.

GENDER DIFFERENCES

In 1973, the National Science Foundation (NSF) began collecting data on the gender mix in academic STEM (science, technology, engineering, and math) faculty across the country. That year, when Marlon Brando won the Best Actor Academy Award for *The Godfather* and Liza Minnelli won Best Actress for *Cabaret*, women made up roughly 10% of America's 118,000 scientists. By 2014, women represented approximately 30% of all STEM scientists. In 2019, the NSF's National Center for Science and Engineering Statistics released its *Women, Minorities, and Persons with Disabilities in Science and*

Engineering report. It contained many important findings with respect to gender differences in science. For example, women held a majority of the degrees (at all levels) in psychology (75%) and the biological sciences (more than 50%). In the social science fields, women earned nearly half of all the degrees awarded in 2016, except in economics. In the past two decades, however, the share of women obtaining bachelor's degrees in mathematics and statistics has declined. Despite increases in the number of women receiving computer science degrees over the past 20 years, computer sciences have one of the lowest percentages of female degree recipients among the science fields. Although women have reached some equality with men in the numbers studying science and engineering in college, with half of the bachelor's degrees in these fields awarded to women in 2016, they still remain underrepresented in these actual occupations. Thus, while women have made headway in becoming researchers in a broader set of fields, they still have a ways to go, along with transgendered and binary persons.

These diversity changes in science have led to shifts in the composition of and dynamics within research teams, with debates on what this means for collaboration. Several empirical studies have suggested that gender-balanced teams are more effective and productive. A recent one has a title that says it all: "Gender-diverse teams produce more novel and higher-impact scientific ideas." It found that teams made up of both men and women produced more innovative and frequently cited papers than those with all-men or all-women teams. The greater the team's gender balance, the more these advantages increased, and this appeared to be nearly universal in all science fields.

Other studies have shown that gender-based conflict sometimes occurs in mixed-gender research teams. Drs. Monica Gaughan and Barry Bozeman found that both status and gender were used as interpretive frames for collaborative behavior by scientists in research teams, with more emphasis placed on status hierarchy than on gender differences. While there may not yet be consensus on the impacts—which may vary with field, generation, and other confounding factors—we can agree that communicating and managing projects may be different for mixed-gender versus single-gender teams. Thankfully, more resources are being developed to teach researchers how to manage and prevent conflict in science teams, such as the framework developed by Drs. Michelle Bennett and Howard Gadlin. More researchers are also starting to investigate the inner workings of research teams. Recently, Dr. Hannah Love and her colleagues tested five hypotheses about the role of women on university-based science teams. Delving into details of how the women impacted these teams, the authors concluded with a recommendation: "In the future, when scientists ask 'What proportion of women is ideal on a team?' consider responding with 'It is not about the number of women, but rather how women on teams are integrated and empowered.'"

Over the last 30 years there has been a lot of research on understanding differences in scientific productivity between males and females. A study by Drs. Svein Kyvik and Mari Teigen in 1996 found that childcare responsibilities and a lack of research collaboration were the two factors that caused significant gender differences in scientific publishing. Women with young children and women who did not collaborate in research projects with other scientists were less productive than either their male colleagues or their fellow female workers without such limitations. The burden of family responsibilities for women was exacerbated during the pandemic. If all researchers would strive to learn how to better communicate and collaborate, it would help ensure that more women could lean in with respect to both mixed-gender and all-female working groups.

MULTICULTURALISM AND INTERNATIONAL COLLABORATION

Throughout history, scientists who emigrated from their home countries have played critical roles shaping breakthroughs and enhancing transformational science. Immigrants continue to play an important role in science globally. Between 2003 and 2013, the number of researchers in the United States increased from 21.6 million to 29 million. Over that same period, the number of immigrant scientists and engineers went from 3.4 million to 5.2 million. This increase has led to more diverse collaborations, which bring together more creative ideas and solutions to problems. Cultural differences in how women are viewed or how opinions are expressed, however, can vary dramatically, and this affects how teams collaborate. Often immigrants may feel isolated or think they don't truly belong to teams, due to cultural differences and biases. Just as there can be unconscious biases when it comes to hiring, promotions, and manuscript and proposal reviews, there can also be unconscious or conscious biases against foreign-born colleagues.

Sometimes this may be unintentionally hurtful, such as not inviting a Muslim colleague out for beers with the team. Being aware of how our beliefs about cultural differences shape our own behaviors is a first step toward more genuine collaborations. Of note, ethnically diverse teams still outperform others, with respect to some science publication metrics. One of the top studies is by Drs. Richard Freeman and Wei Huang, who looked at the ethnic identity of US-based authors of over 2.5 million scientific papers from 1985 to 2008. They found that "persons of similar ethnicity coauthor together more frequently than predicted by their proportion among authors," although such ethnic links were "associated with publication in lower-impact journals and with fewer citations."

The ease of using the internet for communication and free international calls has continued to prompt increases in collaborations that cross national boundaries. The National Science Board's 2018 *Science and Engineering Indicators* report found that between 2006 and 2016, publications with international coauthors grew from 24.9 to 37.2% of all coauthored publications. Many science funding agencies are working to make it easier to provide monies for international collaborations.

In any new field of science, an explosion of terminology designed to measure, track, and otherwise describe that field occurs. The science of team science is no different. "Team science" and "research collaboration" may seem synonymous, but there are nuances between these terms. For example, teams composed of members from different disciplines can be described as undertaking unidisciplinary, multidisciplinary, interdisciplinary, or transdisciplinary research. The committee that contributed to the National Research Council's 2015 report on *Enhancing the Effectiveness of Team Science* defined multidisciplinary research as a form of investigation where each discipline is making separate contributions in an additive way. Interdisciplinary research integrates information to solve problems or expand theories. Transdisciplinary research integrates and transcends previous theories and frameworks that extend past traditional disciplinary boundaries. My coauthors and I called our work on the 1993 hantavirus outbreak team in New Mexico "transdisciplinary research," as it involved many disciplines coming together, using new technologies and applying integrated knowledge to solve a problem and create a new framework for infectious disease outbreak investigations, as well as to understand the ecology of diseases in a One Health context.

ETHICS

If you have been in science longer than a few years, you most likely can think of a situation you witnessed or heard about that seemed or was ethically "off." Maybe it was a graduate student who was sexually harassed, or a scientist being accused of misrepresenting or even making up data. While the foundation of science is based on an assumption of the trustworthiness of the data collected, as well as a rigorous analysis of that data, there can be human bias in or a desire for pure economic or professional gain from the results. Although instances of data fraud in research are rare, they do exist. There are also examples of feuds between scientists that lead one scientist to wrongfully accuse another of data manipulation. A few of the top ethical issues in research collaboration include not crediting coauthors, professors being credited for a student's work, students working without adequate training or safety protections, and misuse of the funding received for projects.

Coauthorship

A major issue in research collaboration centers around authorship. It is generally agreed that the people who make the most contributions to a research project, or to the manuscript describing the work and its results, should be included in the final authorship list. While there is no fixed process for deciding whom to include, the advice I received from a colleague many years ago was to ask the question, "Would this research have been possible in this final form without this person?" This is not a foolproof method, but it can help discern who should be on the author list versus who is only mentioned in the acknowledgments. In the health sciences in particular, there have been reports of authorship issues that range from honorary authorship to ghost authorship. Those who have written articles published in the predatory journals that now abound can hardly be considered to be on a par with those whose articles are peer reviewed. In a 2018 study published in *PLoS One*, the mean number of authors in a selection of health science journals increased by almost 23% from 1995 to 2005. Due to concerns about authorship, many journals now require the specific contributions of each author to be spelled out. The most basic responsibilities regarding authorship include ensuring that an individual listed as a coauthor made some useful contribution to the study, played a role in writing and editing the manuscript, and would be expected to defend the study against criticisms. Contributors who do not meet those criteria should not be listed as authors, but they should be acknowledged. Resources about authorship, like the International Committee of Medical Journal Editors, offer examples of activities that, by themselves, do not qualify a contributor for authorship, such as the acquisition of funding, general administrative support, technical editing, and language editing.

In 1968, Robert Merton posited a theory stating that scientific achievement exhibits a "Matthew effect," where scientists who have previously been successful are more likely to succeed again, which increases their distinction. The theory is based on the biblical verse in Matthew 25:29, which states, "For whoever has will be given more, and they will have an abundance. Whoever does not have, even what they have will be taken from them." The Matthew effect is also referred to as "accumulated advantage," where those who have more are better able to acquire more. Merton's theory looked at the allocation of rewards to scientists, such as authorship and funding, where credit has been disproportionately accorded to more-senior researchers, regardless of their particular contribution to the project. The Matthew effect can also be applied to citations. The person whose article receives citations in other publications will garner more, and the person who has the most citations will accrue them exponentially faster. In the same article, Merton also introduced the concept of the "41st chair." Science has the special problem of being a system where individuals or organizations take on the job of gauging and rewarding lofty performances on behalf of a large community. The example Merton used is the Nobel Prize. A considerable number of scientists who have not received this prize, or probably will not receive it, have contributed as much, or more, to the advancement of science as some of the recipients. The term "41st chair" refers to a practice by the Académie Française, the principal French institution for matters pertaining to that country's language. The Académie, which was created in 1635, "decided early that only a cohort of 40 could qualify as members, and so emerge as immortals." Through the centuries, this limitation has led to the exclusion of many talented individuals who have achieved their own form of immortality. Merton's list of occupants of this 41st chair included "Descartes, Pascal, Molière, Bayle, Rousseau, Saint-Simon, Diderot, Stendahl, Flaubert, Zola, and Proust."

Segregation

In 1971, Thomas Schelling developed a dynamic mathematical model to answer the question of whether racial segregation can still manifest itself in a community where everyone is intent on not living a segregated life. Schelling pointed out that people become segregated every day by gender, race, age, language, culture, beauty, or sexual orientation. In Schelling's fictional dynamic model, which resembled a checkers game, even if every person was interested in living in a diverse neighborhood, the system still resulted in segregation. This offered strong support for the theory that racism, as in Schelling's example, could operate subtly, even if every person was well intended.

What does this dalliance into the sociality of science and humans have to do with research collaborations? In them, we can be segregated by the same things at work as we are in our society. With the Matthew effect, our position in the social network influences our chance of success. Attending an Ivy League university, having a superstar advisor, or being part of an old boys' club all improve your starting position where you enter the game of science. Due to the scarcity of research funding and of the limited number of positions in science, the 41st chair is a real thing. Small positive biases—such as gender, nationality, well-connected advisors, or even luck—feed into larger positive biases down the line. This social system in science is understood either consciously or unconsciously by researchers, and it is why authorship can become a power struggle. The largest randomized controlled trial on publication bias showed that having a big name in science will help get your paper published. In a study led by Jürgen Huber, published in 2022, only 10% of the reviewers of a particular manuscript recommended acceptance when the sole listed author was unknown. Yet 59% of them endorsed the same manuscript when it carried the name of a Nobel laureate.

Thomas Schelling's results—indicating that no matter how hard we may try to level the playing field, segregation still occurs—are somewhat depressing. Still, I am hopeful that we can make a difference at our local level of collaborative teams. In many ways, the Matthew effect is the capital we all gain as we age and become more accomplished. The *Oxford Dictionary* defines "inclusion" as "the practice or policy of providing equal access to opportunities and resources for people who might otherwise be excluded or marginalized." In our research collaborations, this means continuing to think about and address our own unconscious biases. Many argue that academics and science are meritocracies. But meta-analyses, including two recent ones published in 2018, have shown that while many women have strong publication lists, females are more often second authors. Moreover, sociologist Molly King and her colleagues found that male scientists are 56% more likely to cite themselves than women are, with this trend having significantly risen over the last 25 years.

For women, the combination of fewer publications in high-impact journals and fewer citations leads to something that historian Margaret Rossiter called the "Matilda effect," named for Matilda Gage of New York, an American feminist critic and suffragist who, in the late nineteenth century, both experienced and communicated about this phenomenon. Using the many cases of women scientists who have been ignored, been denied credit, or otherwise dropped from sight, Rossiter expanded on Gage's initial concept, extending it to the bias against acknowledging the achievements of women scientists whose work was attributed to their male colleagues. This situation has improved somewhat, but both unconscious and conscious biases still exist in the science workplace.

Another study looked at awards in the science and technology fields in 13 STEM disciplinary societies. Dr. Anne Lincoln and her colleagues found that while professional awards and prizes have increased for women in the past two decades, men continue to win a higher proportion of awards for scholarly research than would be "expected based on their representation in the nomination pool." They also pointed out the ghettoization of women scientists. (This led me to look up the definition of "ghettoizing," which is the segregation or isolation of a group and the placement of that group into a figurative or literal position of little power.) Increases in the number of awards restricted to female recipients have been initiated to combat the biases of science awards. Yet, as the authors pointed out, "Women-only awards can camouflage women's underrepresentation by inflating the number of female award recipients, leading to the impression that no disparities exist." The real question is how women fare in awards meant for all members. For example, examining awards conferred by the American Physical Society, Lincoln and her team found that including ones restricted to women increased the *proportion* of female winners of that society's awards by 55%, although that represented an increase from 4.2% to 6.5% of the *total amount given*.

Other Forms of Bias

There are many examples of gender and other biases in science, and this topic is worthy of a whole book in itself. It is possible, however, to become aware of both conscious and unconscious biases and move toward not being prejudiced against another group of individuals. A central bias that still exists in science is believing that intelligence differs with gender, race, nationality, or other criteria. The Matilda effect is visible in the publication process, stemming from the perception that women are less competent to publish at the outset of their careers and are given less leeway in what they can and should publish throughout their life's work.

Project Implicit is an international collaboration of researchers who have been studying unconscious biases for two decades. They are interested in implicit social cognition—that is, thoughts and feelings outside of one's conscious awareness and control. The project's goal is to "educate the public about hidden biases and to provide a virtual laboratory, to test if a person has any implicit bias." The many tests for this type of bias include skin tone, disability, gender in science, age, sexuality, race, weight, and several others. Out of curiosity, I completed the gender-in-science test, which required sorting, as fast as you can, types of science or humanities fields into their respective categories, as well as placing them into gender categories. My responses on the test suggested that I had little or no automatic association of "female" with "liberal arts" and "male" with "science." That result might have been because I was in the middle of writing a chapter about gender bias in science and had read dozens of studies on the topic. Everyone has both implicit and conscious biases, and I could guess which items in the test category might highlight those biases. In a group or a team, talking about unconscious bias is a good step in the right direction, but talking about conscious bias may also be divisive if it is not done in the right way.

In terms of an implicit bias for gender, there is a classic riddle that highlights what can lie hidden in our subconscious. A father and his son are in a terrible car crash. The father dies at the scene. His son, in critical condition, is rushed to the hospital and taken into the operating room. When he is about to go under the knife, the surgeon says, "I can't operate on this boy—he's my son!" The reader (or audience) is then asked how that's possible. Responses have included scenarios such as two gay fathers, or one father and one priest (a religious father). But an obvious answer that most people miss when they hear this story is that the surgeon is the boy's mother.

With an increase in international collaboration and the migration of scientists from their home countries, biases against nationalities, cultures, and race can occur. These prejudices against people from other locales and cultures also need to be addressed when discussing diversity and bias in science. In an analysis of 2.5 million publications, a recent study by Michael Bendels and colleagues showed that increased ethnic diversity in an author team correlates with increased citations for the paper and enhances the impact factor of the journal. Research is tackling more complex challenges and requires more diverse approaches and teams to tackle the work. When diversity is lacking in teams, we end up limiting our science.

In a 2018 article, Maria Asplund and Cristin Welle listed a set of actions for every scientist to take in order to combat implicit and conscious biases. The first is to seek diversity. While nearly all midsize to large research organizations, government sponsors, and universities in the United States have diversity initiatives and training sessions, many researchers may only give lip service to the benefits of diversity on collaborative teams. Asplund and Welle stated that since the future of science depends on the highest level of innovation, we have a responsibility to advocate for diversity in our hiring, grant reviews, conference planning, and mentoring. The second action involves making a conscious effort to include those who are underrepresented. Expanding the outreach and not limiting choices only to those who are in the right spot at the right time, and thus are able to get a good start in the game, can help all researchers in the long run. The third action is to share responsibility, meaning that it is everyone's job to promote diversity. It's not the task of the underrepresented members on a search committee to ensure that adequate numbers of women are interviewed; every committee member should do so. The fourth action is to resist and reduce the effects of implicit bias. In recruitment and promotion, we must remember that implicit bias impacts all aspects of our work life, as well as our lives in general. Common evaluation metrics are likely to be contaminated by the biases of a community. The fifth action is to see diversity as an investment in good science. Meeting today's public health challenges, such as global pandemics or climate change, requires rigorous, innovative, team science–driven solutions. Diverse teams produce better solutions, in part because of the challenges that arise from wide-ranging perspectives. By increasing diversity in our labs, departments, committees, and institutions, we strengthen our science.

MANAGING TEAM SCIENCE COLLABORATIONS

Revolutions are defined by change, which can be voluntary or involuntary. Sometimes we actively make alterations in our life, such as moving to a new city or switching jobs. Then there are times when change happens to us. Perhaps a new policy is mandated at work, or a close family member or colleague dies—or a global pandemic hits and disrupts every aspect of life on the planet. Both categories of change can be either positive or negative. An otherwise involuntary change can become an opportunity, and a voluntary move on your part can be a jump toward something better. The changes in science that have happened in the last two decades are both voluntary and involuntary, and I would argue that the majority of them are positive in nature.

One scientific revolution that will be here to stay is the importance of research collaboration in general. Larger, more diverse, multidisciplinary, and international teams are needed to meet complex global and local challenges, as well as to cope with the breadth of research methodologies and advancements. The strategies in this book can be used for teams of 2 or 200.

Roughly 20 years ago, I stopped by a colleague's office at work and was asked if I wanted a cup of coffee. I grabbed a mug from the plethora of more than 20 that were scattered literally everywhere around the office. When asked why there were so many, my colleague responded that, once having calculated the time spent in looking around the office for a mug throughout the day, lots of time would be saved by bringing in a bunch of mugs, so as to always have one readily available. It was forethought I would have never considered until that moment. In my most recent visit to that person's office before the pandemic hit, my colleague commented, "I would ask if you wanted a cup of coffee, but I have no mugs." I looked around to see only one sitting on the desk. I asked where all the mugs went and got the response, "I don't know." (Perhaps those who were offered coffee walked away with the mugs.) I share this story to show not only how we can lose things over time, but also how our individual differences in our approach to work and life are what make collaborations so special. Just like the various ways in which we arrange our physical spaces, we all approach problems and organize our ideas differently. The colleague mentioned above is known for collaborating and including everyone, as well as for offering coffee. The metric for this collaboration could be the number of missing coffee cups.

MANAGEMENT STYLES

In pondering how the scientific revolution has changed or impacted our research collaborations, Bozeman and Youtie developed an ontology of approaches for the management of scientific collaborations. The first, and least favorable, is tyrannical collaboration management. Here, power is exercised in an arbitrary and cruel manner. The next—and most common—approach is directive collaboration management. One person is still in charge of the collaboration and makes the key decisions, but not in a malevolent way. The third approach is pseudo consultative collaboration management, which involves a veneer of democracy despite an evident hierarchy in the team. The fourth style is assumptive collaboration management, where members assume that everyone understands and is in agreement about preferences for the project. The fifth—and most effective—approach is consultative collaboration management, where the team members are consulted at key points over the duration of the collaboration, thus identifying respective preferences and values, in order to decide on specific actions. The strategies and ideas I collect and present in this book are geared toward helping foster consultative collaborative management. While the main idea of this management style is to allow all members to voice their opinions on the major aspects of a collaboration, the primary point is that expectations are understood by everyone on the team. Communication is open, transparent, and frequent.

Understanding complex or "wicked" occurrences—such as climate change, pandemics, and societal issues surrounding disparities—requires the conjunction of a wide range of disciplines. These multidisciplinary teams often include managers from the sponsoring program and other policy stakeholders. In the growing field of the science of team science, there has been mounting evidence that respecting individual attitudes, beliefs, and behaviors makes for successful and effective multidisciplinary research collaborations. I would argue that the dedicated researchers who study scientists are also part of the scientific revolution itself.

BELIEFS

It has often been said that a belief is a thought you think over and over again. The beliefs we have shape our everyday experiences. Our beliefs about how to do science were shaped by our university education, graduate advisors, books we have read, and our first experiences in research. Most of us can think of an advisor who took us under their wing to share with us their personal insights into how to actually do research. We learned strategies for how to be productive each day, how to manage our time, and how to treat our research colleagues and all the stakeholders surrounding us. Our beliefs were shaped both by wonderful examples from our mentors and colleagues, and horrific examples of how not to do things.

We may also have limiting beliefs regarding our own potential and intelligence, which impact how we collaborate. Luckily, beliefs can be questioned, inspected, and replaced by new, better-serving beliefs. When it comes to our own beliefs about collaborations, are they, in general, favorable? Do we think that by participating in a collaboration, we will be more productive, the work will be more innovative, and the experience will generally be enjoyable? How do we view diversity on our teams, and what should the shared values be for the team? And, after pondering all that, how hard do we hold on to our beliefs about collaboration?

COPING WITH THE EFFECTS OF COVID-19

The amazing poet Sonya Renee Taylor used Instagram to share her thoughts about the pandemic, which highlighted not only the impact of the pandemic on us, but also how it brought into focus the things in our life that maybe we did not miss. She wrote, "We will not go back to normal. Normal never was. Our pre-corona existence was never normal other than we normalized greed, inequity, exhaustion, depletion, extraction, disconnection, confusion, rage, hoarding, hate and lack. We should not long to return, my friends. We are being given the opportunity to stitch a new garment. One that fits all of humanity and nature."

For the most part, the COVID-19 pandemic was—and, at the time of this writing, continues to be—horrific in almost every way. It did, however, provide opportunities for some individuals in business and in multiple fields of science. While people pivoted to participating in online video meetings and working remotely, several aspects of their jobs either were eliminated or simply stopped seeming necessary. Hands-on research in the laboratory or in the field requires being in that particular place. Thus the ways in which the pandemic has changed research collaborations will lead to expected, unexpected, and as-yet-unseen changes.

In Meetings

A colleague studying how the pandemic has affected scientists and collaborations is asking researchers the following questions. How have your time management and methods of communication changed? What online collaboration management and communication tools are you now using? Researchers are leaning in to tools such as Slack and Mattermost, as well as dealing with the reality of video conferencing fatigue. Gone are the whiteboard-based brainstorming sessions that fueled both ideas and inspiration. In talking about missing this useful tool in one of my own research team's video calls, we discovered the whiteboard application within the video management system. After about 30 minutes of practice, the team used this app and saved our notes. The multiple colors resembled a kindergarten group art project, with bright yellow words at the bottom urging us to "collect data now."

In one 2020 study of over 3 million knowledge workers in 16 cities around the world, the researchers for this Harvard working paper compared the current number of meetings with pre-pandemic levels. They found that in the first year of the pandemic, the number of meetings per person rose by almost 13%, with the number of attendees increasing by over 13% per meeting. The average length of meetings, however, decreased by over 20%. Collectively, people spent less time in meetings per day in the post-lockdown period. The authors also found a feature that is of common concern among my colleagues—the pandemic led to significant increases in the length of the average workday, generally by over 8%.

All of my 10 colleagues, when asked, agreed that videoconference fatigue is a real thing. Many also agreed that they are often frustrated after calls or online conferences. According to Dr. Elizabeth Keating, a linguistic anthropologist at the University of Texas at Austin and author of the book *Words Matter: Communicating Effectively in the New Global Office*, a big contributor to our frustration has to do with nonverbal communication and peripheral participation. Keating pointed out that with in-person meetings, our looking at other people, and seeing them observing us during work hours, creates an important foundation for teamwork. It helps build trust, confirms mutual understanding, and lets us understand the emotional state of the other participants in the meeting. If someone is discouraged or depressed, it is easier to become aware of that in an in-person meeting, as opposed to that individual shutting off their camera and not speaking during the call.

When one of our coworkers seems down, we can check in, ask how they are, and perhaps suggest sharing a cup of coffee or going out to lunch. Dozens of books on nonverbal communication have been written, and we can see the impact of losing that ability in our virtual team meetings.

The second important part of our daily communication routine at work has to do with our "peripheral participation" and how new people can feel accepted. Peripheral participation interactions happen in the breakrooms, on the trips down to the cafeteria for lunch, and on afternoon walks around the building or campus to shake off sluggishness. During these short and relatively private activities, we can ask our colleagues about what happened in meetings or discuss our interpretation of behaviors. As Keating pointed out, when we have video meetings, we miss so many nonverbal communications and cues that can build trust. Moments of true connection are lost. Once, when I was leaving a meeting, my colleague, who was a behavioral primatologist, turned to me and said, "John sure was embarrassed when that topic came up." When I asked what generated that comment, the response was, "His ears turned bright red." My colleague also commented that when John was nervous, he would pick up his pen, smell it, and then set it down again.

When we hold meetings in person, we lean in, put a hand on a shoulder, cross or uncross our arms, and smile or frown to help communicate how we are feeling in general, or toward another person. As we shifted away from in-person meetings, emails have taken on even more importance as the predominant mode of communication. As Erica Dhawan, author of a 2021 book on digital body language and connectional intelligence, stated, "A phone or video call is worth a thousand emails." Although "Uh, you're on mute" was the most used sentence in 2020 (and in 2021, 2022 . . .).

By Modes of Behavior

For many of us, our beliefs and behaviors while working collaboratively have become more entrenched over time. We have developed our own set of rules for communicating, setting boundaries, not being a doormat, and pulling our own weight. With the scientific revolution there does seem to be a greater acceptance of diversity in research and a kinder, gentler mentorship. There is a current movement for a kinder approach to science, using the definition that "kindness" means protecting and promoting the well-being of others. Yet how do young researchers best prepare for this new scientific revolution and for more effective collaborative work? The first answer to this question is to focus on learning about collaborative techniques and relationships at work. Many decades ago, learning strategies—such as effective communication, active listening, the different aspects of building trust, and the development of collective values for teams—were not part of the curricula in most science fields. While 75% of all communication is nonverbal, 70% of communication these days, even before the pandemic, is virtual. Learning how to effectively communicate is a critical skill for all scientists. In our classes we are guided through syllabi on the technical aspects and theories of our chosen fields, but we muddle our way through methods of working with others, attempting to communicate, and otherwise trying to be understood.

Digital body language is the next frontier. The most common medium for this is email. Every one of us has a horror story, which we can recount in gory detail, about a miscommunication through an email or a text message. We may feel slighted when we stay up late to finish a detailed analysis or write a section in a long technical proposal and the only feedback we get is either "Looks good!" or "This could use some work." Even worse is writing a detailed email with a question for a colleague but never getting a reply, a situation known as the "big black hole." While most of us learn the hard way—once—not to hit "Reply all" for an email containing some personal information, we do not have simple strategies for writing effective emails to our research colleagues. According to Dhawan, there are a few questions we can ask ourselves to help with the digital body language of email. Did I give enough context? Is it clear what I mean? Am I using the right emotional tone? How can I use punctuation to show emotion in the right way? Am I showing a clear call to action or a direction to take? Never confuse a brief message with a clear message. We can work toward creating a culture of clarity on our research teams in many ways. Rereading our emails and editing them can answer several questions. Am I being clear about what I need? Did I send it to the right people on the team? Am I intentional and realistic in my expectations and deadlines? We all want to be productive and get the most done as quickly as possible, but not rereading messages and checking for key points can create mistakes that involve more time to clean up than it would take to more consciously communicate the initial message. Recently I wrote two emails with a request and listed the wrong due date—one week after a proposal package was due. It is also important to remember that while brevity may be convenient for you, it can add worry and stress on the receiving end.

I apparently write like a pirate when composing emails. It is not that I am making typos or neglecting to proofread and edit. When I write, I can mentally hear my voice, and my brain skips over small words of three letters or less, such as to, be, at, is, and for. My mental voice makes it seem like they exist in the text. I had a professional science assistant once who would read my emails out loud in her office with a pirate voice. To correct this issue, and avoid the humiliation of my colleagues possibly not taking me seriously, I have the note "Go slow" taped to my computer, to remind myself to slow down when crafting emails. And when I read and reread my emails, I focus on looking for these small words.

Other digital body language strategies can also improve the ways in which we communicate our intentions, feelings, and directions. The world has shifted to using punctuation and emojis to help communicate our authentic selves. As Kathleen Barber has shown us on Twitter, "Every email I ever send: Hello! I am extremely excited to be corresponding with you! You can tell by the number of exclamations points I use! Here is a sentence with a period so that I don't come across as manic. Thanks!" Most of us have used these methods in our emails, particularly when we are trying to gain familiarity with our team members. Using the right medium that matches the length, complexity, and familiarity of our message can go a long way toward gaining trust within our collaborations.

Power dynamics can play in a role in our personal relationships, but in science, our colleagues may often be our peers. How do we get our colleagues to lean in to getting the work done, without resorting to outside authorities? Often that question comes down to how much we trust each other.

Through Various Types of Intelligence

The foundation of science is traditional logical-mathematical intelligence. Scientists, engineers, and other people in technical fields must be able to move through an educational curriculum based on mathematics and statistics, as well as other aspects of research that demand a significant amount of brain power. Yet we all have different combinations of mental and emotional strengths that make us unique. Knowing our own strengths and weaknesses when working with others is important in and of itself. This is also why diversity in our team efforts can produce incredible results, with everyone bringing their unique perspectives and a wide array of talents to the table. One of the best exercises for teams is for everyone to go around and talk about the strengths and weaknesses they bring to a team.

Another type of intelligence that is important for collaborations is emotional intelligence, or the ability to manage one's emotions. Researchers Peter Salovey and John Mayer coined the term in 1990, describing it as "the subset of social intelligence that involves the ability to track one's own and others' feelings and emotions, to discriminate among them and to use this information to guide one's thinking and actions." Much of the present book discusses aspects of how to strengthen our emotional intelligence. A related type of intelligence is referred to as interpersonal intelligence, which is the ability to sense another person's feelings and motives.

The newest type of intelligence now recognized as critical for achieving transformational success is "connectional intelligence," defined as "the ability to create and drive greater value by harnessing your networks and relationships." In many ways it's not *who you know* that is important, it's *how you engage with* the people you know. The colleagues that make for great team members are ones who have the five *C*s of connectional intelligence: curiosity, combination, community, courage, and combustion. Organizations and teams enhance their potential by building connectional intelligence into their culture. It is about asking ourselves, "How do we create the conditions that foster collaboration, innovation, and productivity?"

According to Dhawan, there are three types of connectionally intelligent people: thinkers, enablers, and connection executors. Thinkers spark and generate ideas, and most scientists would fall into this category. Enablers create the structures and forces to get things done. Connection executors mobilize the people and resources needed to get things done. While there are additional groups within each category, all three types of connectionally intelligent people have varying degrees of the five *C*s. Successful teams are composed of a diversity of people who have attributes of connectional intelligence.

SUCCESSFUL TEAMS AVOID OVERTHINKING

Just as we do as individuals, we can also overthink as teams. Overthinking is when our thought processes get in the way of success. They slow down progress and muddy the water. Five overthinkers together in a room are like a group of golden retrievers swimming in a shallow muddy pond: things tend to get messy. Personality differences can be measured by tests such as the Myers-Briggs Type Indicator. And we know there are analytical thinkers who like to extensively research questions before making decisions. Much of our overthinking comes from soundtracks that play over and over in our heads. One of them may say that we are not good enough in some way, or another may emphasize that we always struggle with talking to people on the phone. Like individual soundtracks, teams can take on soundtracks that might misrepresent the truth of a situation. This mental aspect might even be harder to overcome, because there is agreement in the soundtrack that solidifies our false belief.

Recently, a large research team in which I am a co-principal investigator had to write a few subcontracts for collaborators at various universities. This task was not completely overwhelming in itself, but each of us either had some experience with or had heard horror stories about getting a subcontract approved. The notion that the process was going to be difficult and filled with agony kept playing in our heads like a broken soundtrack. We therefore validated each other in our negative belief and procrastinated, dragging our feet and trying to delay the pain of inevitable hours to be spent filling out forms. We joked that we were glad that science was not as hard as contracting. After a while, we bit the bullet and finally called our contracting office to get the process started. While there were many forms, filling them out was not that painful, and it took a fraction of the time we assumed it would consume. As Michel de Montaigne remarked in the 1500s, "He who fears he shall suffer, already suffers what he fears." All of us can recall something we thought was going to be difficult that only ended up taking a few minutes of our time.

Having a team with a foundation of trust in which anyone can ask, "Is that true?" is critical. Broken negative soundtracks need to be replaced with positive ones that move the team forward. If we have a soundtrack that is not hindering us, but instead inspires us to a higher potential, then that soundtrack does not need to be supplanted. An example of this might be team members agreeing that they can get the experiment done in time to get data to submit for an upcoming conference's abstract deadline.

As part of the most recent scientific revolution, the soundtracks playing for teams have changed, and we can accomplish a great deal more, much faster. The computational power and technologies that have been developed in the last 25 years have shifted how researchers do their work and opened new worlds for them. No longer do we need to print off a figure for a poster and glue it to colored cardboard or send our slides off to be developed days ahead of time—sometimes having them processed at the last minute in the city where a conference is located. With each passing year, I imagine the soundtrack of "That's impossible" gets thrown away and replaced with "That should not be a problem." Except when it comes to subcontracts.

Exercise

Here are some questions to ask yourself and to think about. You might want to try writing down your thoughts in a journal.

1. What differences have you seen over the years between various collaborations?
2. What challenges have you had in your own collaborations and what were the lessons learned?
3. Can you think of an important research advisor you have had in your life? What lessons did you learn from them?
4. Do you have certain types of collaborative teams you prefer over others? Are they small or big, focused or complex?

CHAPTER 4

THE SCIENCE OF TEAM SCIENCE

I've never had a longtime collaborator I would not want over for dinner.

MAC HYMAN, *mathematician*

It is natural for researchers who participated in more than one collaboration to compare the projects. Which teams worked better, and which project was more enjoyable to participate in? We are also either consciously or unconsciously asking ourselves questions about our collaborations. Were the project's team meetings pleasant, or a source of dread? Were there one or two people on the team whom you hated working with and avoided as much as possible? Did the team ever enjoy social outings, or was it all business? Did you procrastinate more, or less, on one of the projects? If the former, what led to you using up more time? Did you feel valued and listened to, or did you feel like a third wheel? What were your pet peeves about your collaborations? Were you energized enough to continue working with the team on the proposal? What were the mistakes you—or the team—made? How did the team handle failure and challenges? Would you do it all over again?

The field of the science of team science has emerged from questions such as these, and it has helped identify many of the elements in what makes for good and bad collaborations. According to a book by Barry Bozeman, one of the top leaders in team science, and Jan Youtie, there are four basic types of research collaborations—nightmare, routinely bad, routinely good, and dream teams—with the vast majority falling into the routinely good category.

These coauthors have created a schema to help represent the probable frequency of collaboration types (figure 3). In routinely good collaborations, researchers received some career benefit, the team successfully tested the hypotheses, and the deliverables were more or less successfully completed. There was not too much drama, with no one throwing a vase or a bowl of spaghetti napolitana at anyone. The team members would work together again, if permitted, and a few might be Facebook friends. In the current era of video conferencing, enlarged upon by the pandemic, the team might even know the names of their colleagues' pets or small children.

In routinely bad collaborations, many of the participants might wince a little if it was suggested that they work together again. There may have been a temper tantrum or two. Perhaps one or two members often clashed, or one person brought drama and manipulation into the project. Maybe someone threw cheesecake. The project probably still delivered results and produced publications, but a good time was not had by all.

Nightmare collaborations involve restraining orders and lawsuits, and they often enter into departmental lore. Such projects usually end with significant personnel changes or result in people leaving research entirely. Often, there might be extreme clashes between individuals on the team, or perhaps unethical behavior took place. Nightmare collaborations are stressful for all those who were involved and make everyone wonder why they

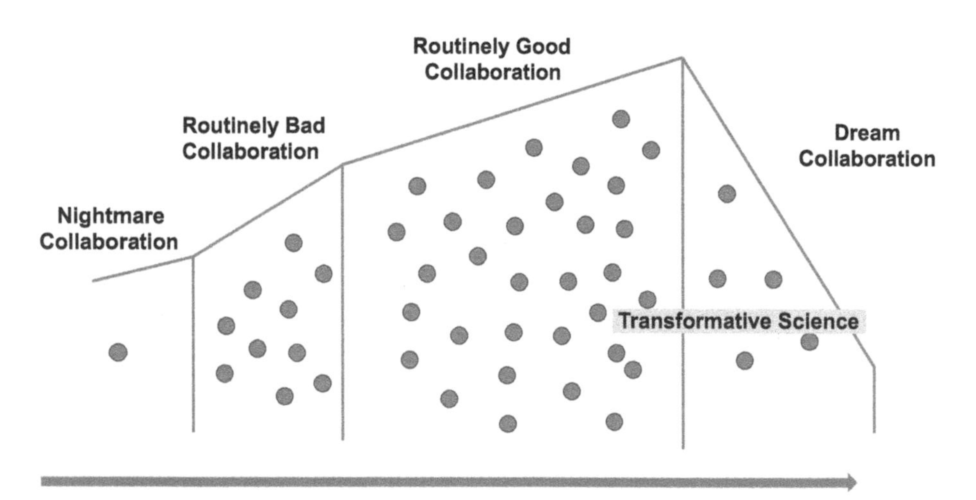

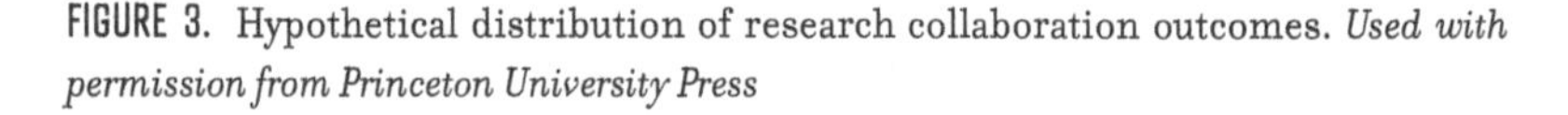

FIGURE 3. Hypothetical distribution of research collaboration outcomes. *Used with permission from Princeton University Press*

chose science as a career. Luckily, as Bozeman and Youtie pointed out, most researchers never experience a single nightmare collaboration.

We sometimes envision dream team collaborations to be like the 1992 US men's Olympic basketball team: filled with all-star and future Hall of Fame players who captured the gold medal. Yet a collaboration in business or science can also be a dream team. We don't have to have a Michael Jordan or Larry Bird of science on our team, however. The 1993 hantavirus team in New Mexico was a dream team, which had many firsts that led to transformational science. While dream collaborations may be uncommon, they are experienced by most of us at some point. Dream teams are highly successful; they experience breakthrough results and publications, great social interactions, and a great degree of satisfaction. If younger students or postdocs are part of the team, dream collaborations may also lead to a person choosing research as a career. While it may seem like a dream team comes together organically, it actually depends on the timing, the scientific question being examined, and funding. The strategies discussed in this book can help make collaborations more rewarding and elevate a routinely good collaboration into a dream team.

The complexities of science, and the many types of collaborations, make a scientific investigation of research teams challenging. The field of team science is defined by an aim for a better understanding of the circumstances that either facilitate or hinder effective team-based research. As Dr. Kara Hall, another leader in the science of team science, and her colleagues pointed out, much of the focus in team science research has to do with the value of team science. This research also looks at the formation of science teams, team composition and its influence on performance, the processes that are effective in the team's functioning, and institutional influence on team science. One of the major impetuses for studying team science comes from funding agencies, which are interested in increasing their return on investment in research. Scientific leaders have expressed concern about why team science is not happening effectively. Both they and scientists in general want to understand the barriers to conducting team science.

For researchers and others interested in team science, check out the International Network for the Science of Team Science. This society has created a Mendeley database of empirical literature on team science and scientific collaboration that has over 3,000 references. Thanks to the hard work of these dedicated researchers in the field of team science and research collaboration, there are many takeaways that can help anyone involved in collaborations or on teams, not just those in science. One of the first points is that *cooperation* is not the same as *collaboration*. Cooperation merely requires a synchronization of people or tasks. Scientific collaboration, as defined by Dr. Diane Sonnenwald in 2008, is "the interaction taking place within a social context among two or more scientists that facilitates the sharing of meaning and the completion of tasks with respect to a mutually shared, superordinate goal."

Thus far I have used the words "teams" and "collaboration" interchangeably, although they are different. Collaboration means two or more researchers working together, but they may or may not make up a team. A team is usually built around a project. In addition, a larger community of people supports the structure of a collaboration or project. These people are the administrative assistants, budget analysts, students, managers, and laboratory janitors. As Keith Ferrazzi and Noel Weyrich pointed out, we can also think of this support community as being on our team. Forgetting about such people or treating them poorly can diminish the effectiveness of the work.

My review of the field of the science of team science in this chapter is by no means meant to be comprehensive. But there are some key takeaways that we can use to increase the effectiveness of our collaborations and teams. There are four basic stages of scientific collaboration: foundation, formulation, sustainment, and conclusion. In the foundation stage, there are drivers, or reasons, for developing the ideas and writing a proposal or beginning a project. These factors include, but may be not limited to, scientific, political, or socioeconomic issues; resource accessibility; social networks; and personal choices. Many collaborations that could otherwise be good ones may never make it out of this foundation stage. That is, a collaboration might be formed to write a proposal that may never get funded. There may be some initial development of preliminary data for the proposed project, but the larger endeavor never takes off. These collaborations may dissolve over time, or they can evolve into the team working on newer ideas that do get supported.

In bringing together a team, researchers tend to collaborate with members of their own institutions. This may be due to distance factors and the ease of working together with familiar individuals. Yet research shows that collaborations crossing geographic and organizational boundaries—known as boundary-spanning teams—have much greater productivity and scientific impact when compared with less broadly distributed teams or solo investigators.

Another theme in the science of team science is the value of cross-disciplinary integration in teams, which can help sustain them. Such teams work together to solve a scientific problem or challenge by integrating concepts, theories, approaches, or methods across more than one discipline. For example, the 1993 New Mexico hantavirus group started out as a medical and infectious disease team. It then grew to include forensic pathologists, small-mammal ecologists, molecular biologists, aerosolization researchers, climatologists, and sociologists. As for the conclusion stage, it has generally been shown that research teams produce more publications than solo authors. Other studies have found a curvilinear relationship between the degree of scientific heterogeneity and research impact, with moderate heterogeneity yielding the most influential publications. For example, Dr. Alfredo Yegros-Yegros and colleagues noted that combining multiple fields had a positive effect on knowledge creation, but research success was better achieved through investigations with a relatively proximal range of fields, as interdisciplinary research in more-distant disciplines might be too risky.

WE ARE MORE CONNECTED

One of the early, and instructive, findings on the patterns of scientific collaboration came from Dr. Mark Newman, a professor and researcher on the structure and function of networks, in a study published in 2004. Mapping the coauthorship of articles in different fields, and following the lead of previous research into coauthorships, Newman used this factor to provide a window into the patterns of collaboration within the research community. Coauthorship of a paper was seen as a way to document a collaboration between two or more authors, which formed a "coauthorship network." While the number of papers per author was similar across the disciplines of biology, physics, and mathematics, the number of authors per paper varied among the subjects studied. Biology had the most coauthors, and mathematics the least. Newman reflected that this could be due to differences in the way research is conducted in these fields. Biological research is often performed by large groups of laboratory or field scientists. Mathematics is more theoretical, with individuals working alone or in pairs. In a 2001 paper, Newman also investigated the "small network" phenomenon, where scientists are more likely to collaborate and coauthor a paper if they have a previous coauthor in common. Newman, using several databases that included over 1 million biomedical scientists, found that the typical distance between any two randomly selected scientists was approximately six links—in other words, there were six degrees of separation, as there are in the larger world of human acquaintances.

The current interconnectedness of the world and the development of social media sites for business and work, such as LinkedIn, point to shrinkage in the degrees of separation. A 2013 study of the network of hundreds of millions of Facebook users, having almost 70 billion relationship links, showed an average distance of 3.9 degrees of separation. Coauthorship is a more structured dynamic, but the connection to scientists anywhere in the world is probably less than four degrees. The take-home message from this investigation of science networks is that it is easier today to globally connect with any scientist, in order to reach out and ask questions or create a relationship that can develop into a collaboration. This type of outreach—being formally introduced to a person via someone you know—is easier now via sites like LinkedIn. Such introductions go a long way in establishing initial contacts. This is particularly true for younger researchers starting to build their social network and reach out to potential collaborators. In science, "cold call" emails, based on specifics about publications or exploring a potential collaboration, do work. Nonetheless, the email initiator should have some familiarity with the recipient researcher's work, and it helps to have a specific intention or question in the message. No person wants someone to merely pick their brain.

INTEGRATION AND TEAM COMPOSITION

The previous chapter discussed the past few decades' scientific revolution, focusing on increasing team size, diversity, and the breadth of the research. But what has this meant for research? Some evidence points to a positive relationship between team size and its resultant productivity and impact. Although research from the science of team science suggests that there is an ideal team size of six to nine participants, Kara Hall and her colleagues reviewed findings reporting that an ideal team size varies based on a variety of factors, including the disciplines involved and the type of scientific question asked. For the hantavirus team, as the relevant questions evolved over time, so did the need for including new disciplines and team members. Another revealing finding about the science of team science was that small teams were more likely to generate new and potentially disruptive ideas. Large teams were more likely to further develop previous breakthroughs or do transformative science. For the hantavirus team, while there were many firsts, such as molecular characterization being used in an outbreak investigation, the team's ideas themselves were not disruptive, but the overall investigation became transformational.

There are documented advantages to having diversity among members of a collaboration. The literature also suggests, however, that too much diversity on a team can lead to fragmentation and inefficiencies that undermine scientific outcomes. Like many things in life, researchers have found that team diversity displays a bell-shaped curve. Moderate levels of *cultural* diversity produce greater-impact publications than teams with either none or very high levels of diversity. The impact of *gender* diversity on teams is one of the most common questions asked by researchers. Findings in articles by Bozeman and various coauthors have been that gender diversity is associated with better end results, but some of the data on this topic is ambiguous, as well as confounded with either unconscious or conscious gender bias. And collaborations in many fields contain few to no women. For example, Erin Leahey found that grant proposals having at least one female collaborator are more likely to be funded, yet the gender composition of teams does not appear to be associated with productivity. A 2018 article on the gender gap also found that women engage in more collaborations than men, although this, too, will vary among fields and with the types of collaboration. For example, women are less likely than men to collaborate with international and industry partners.

HOW DO WE FORM RESEARCH TEAMS?

One of the most common questions by both scientists and the researchers who study scientists is, "How should collaboration teams be put together?" Research on collaborations in science and in business (such as the latter by Hein, Hughes, and Golant) has found that physical proximity is the biggest determinant for two or more people working together. Face-to-face interactions and time spent together is an important part of all relationship-building. The importance of face-to-face meetings seems self-evident, but there are also studies to back up our intuition. For instance, Drs. M. Mahdi Roghanizad and Vanessa Bohns showed that face-to-face requests are 34 times more effective than those sent by email. During the pandemic, Zoom meetings have allowed us to see each other through video, but they have not replaced opportunities to grab a cup of coffee or eat lunch together.

Being in close physical proximity and having occasions in which to inquire about colleagues' research or ask questions after a departmental seminar are salient in the creation of new collaborations. Similar interests in research can lead to the formation of a team, and relationship ties from past successes together can help sustain teams. Networking studies, such as those by Newman, have also shown that the strength of such ties can negatively affect team success, which suggests that a balance of both strong and weak ties is important. If your collaborator comes too close to your own interests, skills, and experience, then new ideas are not forged as easily.

I've found that when large teams are created—such as ones that bring sizeable institutions together to collaborate in response to a call for proposals with the primary objective of building cross-institutional teams—the proposal process can be stressful and more difficult. Often, managers are the ones putting the teams together, thinking that they are creating successful groups. But there is little to no thought given to the personalities of the participants or the diversity of the team. When teams are not formed organically, the top individuals will dominate; younger professionals, introverts, or women are less likely to speak up; and good ideas are sometimes ignored. Time is often the dominant constraint in responding to requests for proposals, but if it is at all possible, time invested in getting to know each other would be time well spent.

COLLABORATIVE INERTIA

Inertia describes an object that stays in its current state without external force being applied. We all feel inertia when we are tired and lack motivation to work on something, whether it is cleaning out the garage, deciding what to do with a collection of vacuums, or writing a promised review for a journal. Collaborative inertia is the phenomenon that describes joint efforts that make slow progress, or those that ultimately do not achieve anything. Even successful outcomes involve frustrations and hard work. Collaborative inertia also refers to researchers who keep working with the same individuals over and over again. Projects with collaborative inertia would fall into the lower half of Bozeman and Youtie's schema of types of collaborations (figure 2). Yet, if collaborative advantage is the goal for researchers seeking such group arrangements, why is collaborative inertia so often the outcome? As previously discussed, studies have shown that collaborative diversity and the strength of the resultant collaboration have had positive effects on scientific output, while collaborative inertia has had negative effects. Collaborative inertia happens frequently in teams and was well defined by Drs. Chris Huxham and Siv Vangen who described this as occurring when the output from a collaborative arrangement is negligible, the rate of output is extremely slow, or stories of pain and a hard grind are integral to the successes achieved. One recommendation is to continuously seek out new partners for collaboration while maintaining relationships with strong collaborators you enjoy working with and with whom you have had successful partnerships. Collaborative advantage means "something is achieved that could not have been achieved without the collaboration."

WHAT CONSTITUTES THE MOST EFFECTIVE COLLABORATIONS?

There is a great body of research and literature on what makes teams, including research teams, effective. One of the models to look at team effectiveness, described by Bozeman and Youtie in their book and in many research papers, examines inputs, processes, and outputs. Inputs include the makeup and design of the team, as well as the nature of the scientific problem the members are working on together. The processes of a team include their communication styles and the cohesion of the group. The majority of the strategies discussed in this book have to do with the processes of a team. Better processes lead to better results, such as funded proposals, successful experiments, and publications. Teams have been thought of as information-processing systems that use their collective cognition toward task-driven results. Or, as the inspirational poster says, "We work better together."

Another process, team climate, represents the shared ideas, perceptions, beliefs, and behaviors of a team. The culture of a team may be representative of the overall culture of the institution. It may also have its own culture, based on the individuals in the team. Team culture has been considered extensively in businesses and organizations, and it has been shown to be closely related to work outcomes. As they say, "Culture eats strategy for lunch."

A third process concerns values about safety, which be demonstrated and communicated by institutional and team leadership. Teams can also create their own bubble of culture that values respect, open communication, and trust. When risk-taking and learning from mistakes are valued, and genial confrontations occur, rather than incidents berating individuals and using shame, these indicate good team psychological safety. Research on psychological safety in teams, such as a 1999 article by Amy Edmondson, has found inclusion and learning from errors to be particularly important for fields that prize innovation, including science. Creating an atmosphere of psychological safety is critical for building the foundations for transdisciplinary research. When talking with researchers from other disciplines, it is easy to make a mistake in understanding what they are saying. The ability to admit your lack of knowledge in a field grants you the freedom to communicate more openly and nonjudgmentally, and to not feel like you're stupid.

TEAM COHESION

We might not think of team cohesion in a collaboration on understanding yellow-bellied sapsucker foraging preferences. But consider instead the ability of Naval Special Warfare's (NSW's) primary unit of action, the troop, to act as a cohesive body, which is directly related to its ability to perform effectively in any operational or combative environment. A master's thesis research project, entitled "Attacking the lion: A study of cohesion in naval special warfare operational units," demonstrated that cohesion in NSW troops is higher than in most teams. (As an aside, what a great title! Maybe I will do something similar for my next paper, "Slaying the dragon: Mountain chickadees' reproductive success during drought.") The study's authors also found that expectation management was extremely important, so they created a perception–reality gap model. The main idea behind this gap is that if a communicated direction from leadership comes into conflict with present and powerful influences affecting the primary group—such as training, indoctrination, and historical culture—a rift begins to form between what is perceived and what is real. In the case of scientific research, a perception–reality gap might apply to changes in a laboratory or field technique. Or to a new way of thinking about relevant modeling. Or maybe to how much time and effort getting research approvals and permits will take, in order to have them in place before beginning an experiment. Expectations that vary among members of the group can destroy cohesion.

Why is team cohesion important in a research project? It is often the case that a series of complicated actions need to take place in a specific order for an experiment to work. There is usually a long list of tasks already accomplished before a team sets off into the field or begins to run an experiment. Coordination, cohesion, and shared expectations form the glue that binds things together. Team cohesion has also been defined as "the result of all forces acting on members to remain in the group." This definition can include the commitment of individuals to the team and pride in being part of the collaboration. It has long been understood that the impact of cohesion in sport and military teams is positively related to team effectiveness, and that its effect is stronger when the project's tasks are more interdependent. This is an important consideration, since having a high level of task interdependence is one of the challenges reported for collaborations. It is noteworthy, however, that I could find no real resources or studies on cohesion in science teams. While there has been less research on how to foster cohesion in science teams, many people suggest that a team's makeup and its personalities, as well as the engagement of its leadership, are important for fostering and maintaining cohesion. The most successful teams I have worked on, where I have also had the most fun, were those that had the most cohesion around a specific goal. Going back to the 1993 hantavirus outbreak in New Mexico, team cohesion was extremely high.

GENDER AND DIVERSITY

In doing the background research for this book, I talked to hundreds of scientists. More than 50% of them were women researchers, and all were happy to tell their stories. In fact, many thanked me for "letting [them] be heard." All of the female scientists had specific examples—sometimes small and sometimes big—of being and feeling marginalized in the workplace. While most of the more senior scientists added the caveat that acceptance of women in science is getting better, they all agreed that there is still room for improvement. This was particularly true for scientific collaboration. The range of responses on the differences between how men and women are treated in research collaboration ranged from minor annoyances to major grievances. Almost every woman I talked to could recount stories of being introduced by their first name, while a male colleague was referred to as "Dr. Smith." One primary question we still may have is, "How differently do men and women approach collaboration?" In the emerging scientific revolution, this aspect may also be evolving as researchers adapt and change.

Participating in energizing and productive projects and teams can be great experiences for scientists, but there can also be downsides. One such example is collaboration overload. This can happen when too much effort is taken to make people feel connected and to socialize during all aspects of a project. This is especially true for larger teams that have many multidisciplinary components, with everyone working together. While everyone on a team may experience collaboration overload, women feel this burden more. To support this assertion, studies have also demonstrated that women are more likely to care for the collective team. In a 2016 *Harvard Business Review* article, Rob Cross, Reb Rebele, and Adam Grant published a wonderful summary of collaboration overload and the extra burden carried by women. Science, like business, can be a place where there are tough disputes and harsh arguments over ideas and theories. Research on gender differences in work, such as in *Hardball for Women: Winning at the Game of Business*, has shown that men can have an intense argument—one that borders on throwing punches—and then grab a beer together afterward. Women, in general, are not able to disengage from an argument as quickly and will internalize the discussion more. Another key difference is that women are more likely to agree with the statement "being a good team player means helping all my colleagues with what they need to accomplish." On the other hand, men are more likely to agree with the statement "being a good team player is knowing your position and playing it well." This can have important ramifications for the ability of research teams to work together.

Another reason why women tend to feel more of a collaboration burden is that they are less likely to carve out time during the day to turn off their phone and focus on the important work at hand. Feeling guilty for not being on every committee, volunteering for every departmental function, and being more supportive of their families at home can make it nearly impossible for women, in particular, to shut out the world. Saying "yes" can be so ingrained for women that it almost feels like it is in our DNA. Even with the emerging movement to say "no" more often and take control of our time, this can be a nearly impossible loop to break. There is positive feedback from staying late at work, stepping up and filling a need, or always being the dependable one. Recently, in one of my larger team projects, several of the early-career women researchers pointed out that they felt as though they were asked to take notes more often than the men. Most of us have been in a meeting when the question "Can someone take notes?" arises, followed by a 1-2-3 beat of silence until one person eventually offers to do so. Even with most women thinking, "I will not volunteer quickly to see if a man will offer to take notes," when no one initially responds, it then becomes a chance to save the day and be the hero. This is an opportunity to be helpful, which is a core value held by many of those who are prone to be people-pleasers. In a 2015 *Forbes* article on ending gender bias, Sir Richard Branson, founder of the Virgin Group, pointed out that "99% of people in leadership roles don't take notes." Branson's advice, now shared widely, is for every attendee to take notes during meetings. "Men shouldn't take over the note-taking from women; everyone should be taking notes."

One day an email to everyone in our department came through from a male scientist, asking that the dirty dishes in the communal kitchen be cleaned up. Because my office was right by that area and I needed some tea, I went to the kitchen and was aghast to find a sink full of dirty dishes. Since I was in a half-management, half-science position at the time, we discussed this later that day in a management meeting. I was surprised to hear that the scientist sending the email was not just pointing out this state of affairs for whomever had left the dishes there, but that he was the one who had been putting his dirty dishes in the sink. Cleaning up, like note-taking, is what Adam Grant and Sheryl Sandberg called "office housework," and the key to a successful working environment is to create a culture where everyone contributes to these types of things. Many senior scientists may say that they worked hard to get to where they are, and menial tasks are beneath them. That may be true, but a science team's cohesion and connection come from a mutual respect that is real as well as perceived. Often women and minorities are acutely aware of stereotypes that may be out there about their intellect or work habits. Thus women or minorities may work harder to combat such misconceptions. We may also apply these same stereotypes to ourselves, which can affect our self-confidence.

Scientists will vary with respect to the factors influencing their decisions to collaborate, and much work has focused on the effects of gender on research collaboration. Bozeman and Gaughan sought to answer the question, "How do men and women differ in their research collaboration patterns and strategies?" Their primary hypothesis was that men and women faculty researchers differ significantly in the number of their collaborations and in various other aspects, such as marital and tenure status. Their second hypothesis was that men and women faculty researchers differ significantly in their strategies for choosing collaborators. What Bozeman and Gaughan found was that women have slightly more collaborators than men, and that both men and women affiliated with university research centers have more collaborators. These coauthors were also interested in how men and women differ in their approach to collaborative strategies. They classified these strategies into four categories: instrumental (concern with immediate work factors, including the assignment of credit), experience (previous participation in collaborations), nationalist (a desire for collaborators from one's own nation or who share the same language), and mentoring (a desire to help graduate students and junior faculty). They found that men oriented more toward instrumental and experience strategies, and women were motivated by mentoring ones.

Strategies for collaboration can refer not only to *how* we collaborate, but how we *seek out* and *choose* collaborators. Many researchers might say that they look for collaborations that are either already funded or have a high probability of obtaining funds. There are a variety of strategies at play here, too. Taskmasters tend to prefer collaborators who subscribe to their own work ethics, and they focus more on tasks than on teamwork. Nationalists generally select collaborators who are fluent in their own language, are of the same nationality, or have other similarities. Tacticians usually base their choice of a collaborator on whether or not that person has skills that complement their own capabilities. Mentors are motivated to help junior colleagues and graduate students by collaborating with them. Followers enter into a collaboration because someone in the administration requested that they work with a particular individual, or because the potential collaborator has a strong science reputation. Buddies choose collaborators based on the length of time they have known each other, how much fun they have had working with that person, and the success and quality of a prior collaboration. Several studies by Bozeman and colleagues have found that women have more collaborators, in all forms of collaboration strategies, than men, except when it comes to international collaborations, where men engage with more of them.

RESEARCH COLLABORATION EFFECTIVENESS

By now it should be clear that collaboration increases scientific output and productivity. But since there is a continuum of collaborations, ranging from nightmare teams to dream teams, it can be difficult to see the value of collaborative efforts if you have experienced poor ones. Is the return on investment worth the resources needed to improve the effectiveness of collaborations? Like most fields in science, research has shown that collaborations are complex, and studies often report opposite findings. For example, studies by Bozeman and various colleagues have shown that collaborations between academia and industry often exhibit more conflict. It is good to be aware of these different mindsets between institution types. In addition, academia and industry may have different missions and bottom lines for research. Yet they can also be partners, with a shared vision of solving a problem, and complement each other to jointly accomplish bigger things.

In general, team science researchers have found that the ways in which we work together can affect our ability to do so for an extended period. This includes how we address conflict within the team, our communication styles and networks, our previous collaborations and the overall experience of working together, and the management styles of the leaders. Other factors that can impact the effectiveness of a collaboration include cyber infrastructures for communication and for sharing data and analysis results. Having either an internet platform for communication or a process in place for sharing data can create an atmosphere of responsiveness and effectiveness.

The examples highlighted in this book are meant to show how things can go right, as well as how things can sometimes go wrong. They also point out the challenges inherent in research collaborations. What can be seen from the various studies on the effectiveness of collaborations is that researchers from diverse disciplines can work together and accomplish great things, especially if careful thought is given to how to encourage and sustain the team and support the individuals on it over the long haul.

Exercise

Here are some questions to ask yourself and to think about. You might want to try writing down your thoughts in a journal.

1. Based on the extant research, how can your own team be more effective? What is one thing that your team could implement from the findings in the field of team science?
2. Do you feel like your research, and your ability to work with others, are at their peak of effectiveness?
3. How diverse is your team?
4. How do you feel that diversity—or a lack thereof—impacts the team?

CHAPTER 5
TRUST

When you share power, you accrue power.
WILLIAM PRIEDHORSKY, *program director, Los Alamos National Laboratory*

The more high-performance the team, the more it requires the safety of unconditional trust.
RICHARD FAGERLIN, *author*

One of the most common interview questions, particularly for leadership or management positions, is, "What is a challenge at work or in managing a team that you overcame, and why was it a challenge?" Early in my career, when I practiced for an interview, I couldn't come up with many extreme challenges. Most of the challenges I had heretofore faced may have been significant, but they seemed manageable within the teams.

If I were asked this question today about a challenge I faced at work, I would not talk about an important project that kept running into issues and roadblocks and describe how the team overcame these obstacles. That situation was fun and energizing. Sending the team the music file of "Gonna Fly Now" from *Rocky* and rallying them to step up to meet the challenge was invigorating. Sure, at the time there was stress involved, but it was a positive stress. It gave me a reason for getting up in the morning, fighting the battles, and inspiring people to live to their full potential.

Now I often ask this same question to colleagues, both in leadership positions and not, and the common theme that runs through all their responses is dealing with difficult people at work. I'd have to agree. Initially, I avoided working on teams that included people others found to be negatively challenging. Later on, I was in a leadership position and had to manage one such person. There was no avoiding it, as I could not opt out. I had to set boundaries for myself and others. I had to decide on expectations and try to give constructive feedback without making that person feel like a victim. I then had to try to brush off their impact on my psyche as I left work each day. I needed to reach into my bag for all the good mental health tools I have gathered over the years, which enabled me to deal with days that included an interaction with this difficult person. These tools included exercise, Bikram yoga, an Epsom salt bath, a hike, hugs for my dog, talks with a friend not involved with my work, overeating, reading another self-help psychology book about dealing with difficult people, meditation, and looking at cat videos. Then I would come to the conclusion that these are all things I love to do, and I would resolve to spend more time doing them.

All of this is to say that working with difficult people is just that—difficult. Having toxic individuals on a team can break down trust, not only between that person and the team, but also between other team members. Often we cannot pinpoint what makes a person hard to work with or be around, but sometimes it is still quite clear to everyone involved. Nor does everyone experience the same behavior from such an individual. People may have different faces or personalities, depending on whom they are talking to. Some manage up well and but cannot manage down. In other cases, a person may lack transparency. Such individuals could be information hoarders, or give different pieces of information to the various team members. Sometimes it is just a nebulous feeling you experience when a person smiles and says seemingly benign things, but your inside voice is screaming, "Flee, you fool!" These are all things that break down trust.

Most of us have had the experience of working for a bad boss. In those situations, while life is not the best, we are still learning valuable lessons about how *not* to be a leader once we eventually become one. We can also look for new positions and opportunities. Everyone understands moving away from a horrible boss, but few would understand moving away from a difficult team member.

Leadership books often pose the question, "Which is more important, having the right people on the bus, or a map of where you are going?" The correct answer is having the right people on the bus. When there are no options for avoiding a difficult person on a team—which is more common in science than in business, due to the need for specific technical expertise or someone who has history with the project—it can be imperative to manage that person positively.

Almost every scientist can think of a principal investigator who was a lousy leader, or, even worse, a toxic and caustic person who divided the team at every opportunity. I have seen the divisiveness that can happen in an organization suffering from poor management. This type of person, even though not intentionally, creates feelings of helplessness, anxiety, guilt, and overt hostility in people within a team or organization. Working with stressful people is potentially devastating, whether they are your direct boss, a graduate student working in your laboratory, or a scientific peer collaborating on a project. The impact on one's emotional health cannot be dismissed, and, in a team environment, the consequences can be disastrous. Power dynamics also come into play. One of the worst and most heartbreaking examples is a toxic graduate student advisor, who can drive promising professionals away from science.

An interviewee for this book recounted the story of a principal investigator who epitomized the classic difficult personality. This manager hoarded information from some people on the team while sharing it with others, purposely dividing the group's members. Their work environment was much like the myth of Sisyphus pushing a rock uphill. One person on the team wrote the discussion section for a manuscript 16 times—being told on each occasion that "This is not right yet"—before finally giving up. There was much rejoicing when that principal investigator left the university. The most significant lesson I drew from this story, though, is that black-and-white thinking by a leader creates long-lasting, similar thinking in others. After a few years, this toxic researcher came back to interview for a higher position at the university. The fear, uncertainty, and pain this situation generated were palpable. Rumors started flying faster than a 5G network. Whispers circulated. How was this person able to make the selection cuts? What if that individual got the job? Was this person going to behave for a year and then dig into people? Depending on their loyalties toward the former investigator, people who had been working together peacefully and productively over several years knew they could no longer trust each other if that person was hired.

When there are few options outside of working with difficult people, it can feel hopeless. But it is important to keep striving to build trust within and between your team members. That said, in such situations it is extremely hard to be transparent and share information, knowing it might be used against you sometime in the future. It can feel risky to share your data or samples and possibly never see the same generosity offered back to you or your team. It's even worse if you or your team members are criticized in public or at a team meeting. When there is no option for a disruptive team member to leave, perhaps because they represent another institution in a collaboration, it can be comforting to know that eventually the project will end and everyone will move on to new research efforts. The team will not be a dream team and accomplish as much as it could, however, if it was not built on a foundation of trust, compassion, and support for each other. As one of my close research colleagues who was going through a tough time at work once told me, "I'm just going to put my head down and focus on the science. The science will get me through, so that I have something to think about."

If we think more about what makes a person difficult to work with, all the reasons primarily deal with one thing: trust. Trust goes beyond whether a person lied once or twice. Trust is a multifaceted and dynamic connection between people, and it is often something we think about in our personal relationships. But as scientists, when was the last time you asked yourself, "Whom do I trust or not on my team, and why?"

There are thousands of books published on understanding and building trust. Much of the material can be summed up by a quote attributed to Under Armour CEO Kevin Plank: "Trust is built drop by drop and lost in buckets." Brands in business are all about trust, and Plank's statement came from a February 2014 interview in *USA Today*, in a response to an incident in the Winter Olympics that year. The week before, the highly anticipated Under Armour speed-skating suit had taken the blame for the US speed-skating team's lousy performances at the Olympics. Even though the American team voted to switch suits during the games, it still finished out of the medals. Even though Under Armour did not lie, cheat, or steal, people's trust in the company still hung on a precipice. Not only is the quote by Plank insightful, but the situation from which it arose is also an example of the complexities of trust. Plank understood that brands are based on trust, but, if lost, that trust could be rebuilt over time through commitment.

To most people, trust depends on honesty and integrity—telling the truth and leaving the right impression. Yet there are many other qualities that also go into trust, such as congruence, humility, body language, variability, competence, and courage. Another component of trust is intention.

Dictionaries define intent as a "plan" or "purpose." The old adage says that the road to hell is paved with good intentions, and we all know that, sometimes, poor behavior turns out to be the bad execution of good intent. Moreover, intention itself is composed of motive, an agenda, and behavior. Motives provide the reason for doing something, and they inspire the greatest trust when they show genuine concern for other people, a greater cause, or a desire to make the world better in general. Motivation leads to your agenda, which is what you intend to do or create, based on your motives. Behavior is the manifestation of motives and agendas. The behavior that engenders credibility and inspires trust is acting in the best interests of others. Thus the intention that creates the greatest trust is one that seeks to benefit all stakeholders or colleagues and is done with a genuine desire to build trust with and between other people, so great things can be accomplished.

MODELS OF TRUST

We can all think of someone in our lives whom we implicitly trust, as well as someone we would not even ask to feed our fish while we are away on vacation. What is it about a person that makes us trust them? We somehow know that they have our back, with our best interests in mind. They are happy to see us succeed and empathetic when we fail. They tell us the truth, and when it is a harsh truth to hear, they are firm but compassionate in the telling. They do what they say they will do, and if they need more time, they are up front about the issues and work to find solutions. People we trust bring the skills they possess to the table and are not afraid to share their weaknesses. They apologize when they make a mistake and own up to it if they have created a problem. They do not place the blame on others. People we trust pick us up when we are down and help us realize our true potential.

People we distrust do not tell us the whole story. They leave small pieces out and present only what is to their advantage. They hoard information and forget to include us on emails. No matter how inconsequential the difficulty or how small the mistake, these people blame everyone except themselves. Deadlines are missed, and elaborate excuses are given. Their word is sometimes good, but not always. When they say they will get the necessary data to you tomorrow, in your head you think, "Maybe that is true, but maybe not." In addition to leaving out important information, they may gossip. People we don't trust will not reach out and ask for help, or admit that they don't know something. In team settings, they may conspire to pit team members against each other. We may or may not catch out a person we don't trust in a lie, but once that happens, trust is broken. In cases where none of the above exists, we may not be able to put a finger on the specific reason why we don't trust someone, but we know instinctively that they don't have our back. As Oprah Winfrey has often said, "Beware of people's intentions, because they always have them."

Trust is the glue that leads to a greater connectivity in collaborations. I was part of a research team once where someone from my home institution told our federal science sponsor that I was "just a veterinarian." My best friend is a veterinarian, and I have spent lots of money over the years paying for their services. I have the utmost respect for them, but I am decidedly not a veterinarian. In response, our team had to make modifications in the proposal and spend time and effort to prove that I was indeed a research scientist, not an animal doctor. While team members did not know for sure which person spread this misinformation, we had good idea who it was, and to this day it would be nearly impossible to place a significant amount of trust in this individual. They definitely did not have my or my team's back.

TRUST MODELS

There are several models that have been developed on how to understand and develop trust. One of my favorites is from Richard Fagerlin's book *Trustology*, which uses the image of a three-legged stool. The legs of the trust stool include integrity, competence, and compassion. Another possible leg for the stool that some trust models have included is communication.

Hundreds of books have been published on each of those topics. One of the best books, which has become a phenomenon in business and personal growth circles, is *The Speed of Trust*, by Stephen M. R. Covey. It lists four core elements for credibility that make you believable, both to yourself and to others. Two of these relate to character, and the other two apply to competence. The first core element is integrity, which can be defined as a consistent adherence to moral and ethical principles—that is, the quality of being complete or undivided. The second deals with intent. What are our intentions when working with others? Are they straightforward? Without self-reflection, we may not even realize what our intentions are. The third is capabilities, which Covey defined as the abilities that inspire confidence, skills, talents, expertise, and knowledge. Capabilities also deal with our ability to learn, evolve, and respond to different situations. The fourth element deals with our results—that is, our performance and track record. While we all have a proven track record, it can be either good or bad.

Capabilities and results are the core elements of the competence leg of the trust stool. In science, competence—in the form of technical skills and knowledge—is focused on and developed. It is often not the broken leg in the trust stool for research teams. But that may not always be the case.

Character, consisting of integrity and intent, is another important leg in that stool. Sometimes the betrayal of trust on science teams creates a break that is beyond repair. I knew a research group that worked together as a strong and connected team for five years. It was a collaboration between the research team, who created the initial project and formulated its ideas, and an engineering researcher, who concretely developed those ideas. One day, using email to share a set of project publications with a federal sponsor, the engineering researcher added "a few citations for the list." The research team was surprised when they opened up the publication list and saw work related to the project, but no mention of their names. They went back to their own laboratory notebooks and confirmed that *their* work was what appeared in the publications. Prior to this incident, they had seen a patent application arising from the project, in which they were not included, get approved. They had let that slide, telling themselves, "The patent was really for engineering, after all." After the publication snafu, however, they felt "intense betrayal, sadness, and anger." There were no apologies. The engineer merely responded that the press of affairs had not allowed any time to check the submitted manuscripts for errors. When I heard the story, I wondered if the real reason was that the engineer thought nobody would ever notice. Instead, the situation was sort of like being pregnant; eventually everyone will figure it out. What was the motivation against the inclusion of research team members in the publications? The research team's only ideas about the problem were that the engineer was either afraid of being scooped or had a deep sense of paranoia. In short, other people, including colleagues on the project, weren't trusted.

After the cat was out of bag about the publication betrayal, the research team used email to officially sever their connection with the engineer, making it clear that they wanted nothing to do with this collaborator ever again. The research team went on to hire their own engineering specialists, who would work directly within their organization. While skills and experience were important attributes for job candidates, the decision makers also emphasized character and wanted all team members to be a part of the hiring interviews.

PSYCHOLOGICAL SAFETY

Trust is the foundation of psychological safety. Do I trust my colleagues? Do they have my back and my best interests in mind? Do they respect me? Will they tell me if I have spinach in my teeth before I give a presentation? We can actively work to build trust in our relationships. One means is through developing psychological safety in our interactions. Psychological safety—a concept made popular by the best-selling book *Crucial Conversations*—is the belief that you won't be punished when you make a mistake. Dr. Amy Edmondson first introduced the idea of "team psychological safety" in 1999, defining it as "a shared belief held by members of a team that the team is safe for interpersonal risk taking."

Psychological safety in teams allows moderate risk-taking in communications, an ability to speak your mind without fear, and creativity. In the above example, the project's joint research and engineering teams had worked together for several years and accomplished many great things. The engineering researcher discussed above was known to be an extremely hard worker. If, for a presentation, you wanted a new deck of slides by midnight, you would get them by 11:55 pm. In terms of the trust model, the engineer had extreme competence and was a diligent, hard worker.

Even so, there were red flags. Not long after that person joined the team, two key investigators spoke with the principal investigators about their new team member. Both women confided that they did not trust this new team member, who was a male, even though neither of them had a good reason for feeling this way. Eventually they saw the engineer speak harshly to students and junior researchers. The students voiced concern about working with this person, and they preferred not to do so, if possible. Many of the younger staff members held back from sharing their concerns, not wanting to complain or seem like whiners. It was also hard for the junior researchers to speak up about how the engineer's students were treated, since they were not the students' mentors. Over time, the team members began to run into other people at their institution that refused to work with the engineer, and the trail of bodies from the verbal attacks began to stack up. But this person had a trust stool with a strong competence leg, based on dependability and getting the work done in a timely manner.

By the time the publication betrayal occurred, the team was not surprised, but they were flabbergasted that the deceit had gone that far. There was an ethics investigation, which did not result in much, and the engineer moved on to a new institution. The two women investigators who had initially complained were angry and hurt that they hadn't been listened to, but they wanted to learn from the experience. As one of them remarked to me, "Along with the glory comes the shit." When I asked them both what they could have done differently, they answered, "Trust your gut and intuition and then monitor behaviors closely. Keep records and ask questions."

We can ask two questions about this example. First, why did the difficult situation occur? Second, how do teams keep this from happening to them? In science, while we are expected to collaborate, the metrics we use are at the individual level: how many publications we have; our Hirsch index, or h-index (the productivity and citation impact of our publications); the number of patents we generate; and how much money we bring in. We are expected to build projects over time, and then eventually build programs. We also have commitments beyond being researchers—teaching, committees, reviews, more committees, mentoring, forms, approvals, public events, and a few more committees. The pressure to succeed in science is immense. This is one of the leading reasons why many PhDs—especially women, who feel the strains of both work and family life—do not go into research-only careers. Moreover, in people who come from a place of fear and scarcity, competition can often breed mistrust.

One of the appealing things about trust is that it creates its own reality. When we don't trust someone, we tend to assume the worst instead of the best. When we trust, however, it is easier to assume the best. The Jargon File, started in 1975, is a collection of slang terms used by various subcultures of computer hackers. In the 1990 version of the Jargon File, the term "Hanlon's razor" was introduced, which was a rule of thumb that states, "Never attribute to malice that which can be adequately explained by stupidity." This same concept has been voiced by others throughout history, going back hundreds of years. In 1774, in *The Sorrows of Young Werther*, Goethe wrote, "Understandings and neglect create more confusion in this world than trickery and malice. At any rate, the last two are certainly much less frequent." Hanlon's razor can provide insights when we deal with colleagues, family members, or institutions. The more we dislike or mistrust someone, the more likely we are to attribute their actions to malice. When someone we mistrust makes a mistake, feeling empathy and understanding tends to be our last response.

In 2015, Julia Rozovsky published the results of a large two-year study on team performance, having interviewed 200 Google employees and looked at 250 attributes of over 180 active Google teams. The study revealed that the highest-performing teams had one thing in common: psychological safety. It also identified five key dynamics that set successful teams apart from other teams at Google. Psychological safety was the top indicator. The next indicator was dependability. Can we count on each other to do high quality work on time? The third was structure and clarity. Are there clear goals, roles, and execution plans for the team? The fourth was the meaningfulness of the work. Is the team working on something that is personally important for all the members? The fifth was the impact of that work. Does the team fundamentally believe that the work matters?

RISK

Scientific collaboration is risky. Dr. Karen Frost-Arnold, who studies epistemology and trust, teaches a class at Hobart and William Smith Colleges entitled Trust and Betrayal. She pointed out the risks of collaborative research with our peers. Our collaborators may perform sloppy, wasteful, or fraudulent work, which can damage reputations. This would come under the competence leg of the trust stool. Then there is the sharing of idea, resources, and materials, a part of collaborations that may be taken advantage of by other researchers. Our collaborators may also plagiarize and take credit for ideas and work that are not their own—as we learned in the above example of an untrustworthy principal engineer in a multifaceted research group—or use ideas or materials to complete their own work and scoop the originator. This would be related to the integrity and character legs of the trust stool. One collaborator may speak poorly of others, spread rumors, treat them harshly, or commit microaggressions. This would form the compassion leg of the trust stool. Frost-Arnold laid out two explanations of trust among scientists. The first is premised on the idea that scientists expect each other to be rational, self-interested beings. We trust each other because being trustworthy enhances our self-regard. This may be the karmic approach to "What goes around, comes around." The second explanation is that scientists trust each other on the basis of their moral character.

In reality, collaborations may have a foot in both explanations. One of the difficulties in a reliance on the self-interest model has to do with power, or the hierarchy within collaborations. Researchers may lack the power to influence the behavior of others, or at least feel like it. There is a recent movement in the workplace with the maxim "See something, say something." Educating people about diversity and inclusion, harassment, and microaggression involves taking some of that power back. It may allow younger female or minority graduate students to feel more able to come forward and point out harassment or other mistreatments in the laboratory and the field. Moral trustworthiness is exactly what junior researchers and graduate students may be looking for.

SUCCESS

Every coin has two sides. In this discussion, one side states that trust is a foundation for a strong collaboration. The other side is the belief that trust is not fundamental to scientific collaboration. There are ample examples in science where the research succeeded in spite of distrust. Some might argue that it even succeeded *because* of distrust, due to an environment of skepticism and competition. Others might say that the success of projects is independent of distrust within the team. In 2001, a social science research team, led by Wesley Shrum, argued that the role of trust in science had been greatly exaggerated. In their study of 53 research collaborations, they found that trust was associated not only with the performance of teams, but also with conflict within teams. The study results, however, offered no way of really knowing if conflict increased or decreased trust. On the other hand, if we look at Patrick Lencioni's *The Five Dysfunctions of a Team*, the second of these dysfunctions is fear of conflict. Lencioni stressed that all great relationships, from marriages to teams, need productive conflict in order to thrive, but you need trust, or psychological safety, to feel comfortable in speaking up about disagreements.

We know from the business world that trust is fundamental for teams to achieve their highest potential. David Hull's premise from his book on the process of science was that "whatever is true for people in general, better be true for scientists." For instance, trust was studied in a 2020 article on teams that collaborate on the construction of high-performance buildings—that is, structures that consume a minimal amount of energy. The authors found that a heightened degree of trust leads to an elevated level of collaboration, which results in high-performing teams successfully constructing high-performance buildings.

HOW CAN WE INCREASE TRUST?

If trust reduces the risk for all partners in a project and increases the overall potential of the research, how can we foster trust within research teams? Trust implies positive outcomes, but it also represents vulnerability and risk.

IMMERSIVE TIME TOGETHER

At the beginning of a collaboration, communication is king. In the study of trust within high-performance construction teams, one of the powerful ways of increasing trust was by starting with a two-day seminar. While most projects nowadays have kick-off meetings, having a full two days with all members of the team, possibly at different locations, is almost unheard of. But we should think about adopting this practice. Having two entire days, without interruption, to go through the goals and objectives of a project with everyone involved in it being present, can ensure that all of the expectations are clear for everyone. This extended meeting can include brainstorming on how to deal with the challenges that arise, and all collaborators can ask any questions that come up.

A kick-off retreat is also a good way to increase communication and trust at the beginning of a project. Another option for an immersive experience for teams, one commonly used in business, is the idea of the sprint. This concept started at Google Ventures and was made popular through the book *Sprint*. A sprint is a weeklong process for answering crucial questions around a central issue or topic. All participants clear their schedules for five days to work on the central theme through a structured process, with the goals set and the tasks completed by Friday. Five days is a lot of time, and some might argue that research does not work as well in sprints. Here I lay down the challenge that maybe there are times when having everyone working together on a project in a short sprint can lead to big leaps in productivity and understanding. Not having to use the strict template of Google's sprint, teams could design their own agendas, which could include side-by-side writing time, or perhaps early, coordinated start times for experiments. While the sprint is designed to get a lot done in a week, one of the byproducts is having the team spend time together. Getting to know our colleagues builds trust.

I have taken part in a few sprints, in order to make quick progress on a project. The sprint forced us to all stay focused on the project and not be sidetracked by the millions of other things team members had on their plates. My first sprint was to develop a documentary presentation about a science conference on epigenetics that we held in the Middle East. We took over a conference room with three full walls of whiteboards, which were filled by the end of two days. For me at least, I knew I was hooked, not only on the accomplishments achieved in a short time, but on my enjoyment of the process and the chance to work so closely with my colleagues. My second sprint brought together a big team to focus on the integration of several larger models. This sprint helped keep the momentum going and let us work out problems as a team. Such timely immersive meetings have a high return on investment, despite the commitment to clear schedules for multiple days. This model-integration team now meets regularly for two-hour "synchs" that keep us on the same page.

Scrum, developed by Jeff Sutherland and Ken Schwaber, expands on sprint techniques. Scrum, in this sense, is defined as "a lightweight framework that helps people, teams and organizations generate value through adaptive solutions for complex problems." The term "scrum" was borrowed from rugby, where it means "the ordered formation of players, used to restart play, in which the forwards of a team form up with arms interlocked and heads down, and push forward against a similar group." The idea behind a project-based scrum was to create a process for teams to work through complex problems. One of the primary goals of scrum is transparency, as the work is meant to be visible to everyone within the team. The five values for scrum are commitment, focus, openness, respect, and courage. Some research collaborations may not fit into the structure of a scrum, but many types of projects could take a big leap forward by using these methodologies. Scrum itself has its own lingo, geared toward project management. It has many devotees around the world, and a podcast is available.

CONNECTION

There are family, friends, or colleagues we have a special connection to and who would be considered part of our inner circle. This link might be through shared history, interests, or something we can't quite seem to put our finger on. One of the reasons why sprints and scrums are successful is because time spent working closely together toward a goal creates a shared history and a connection. In his book *Integrity*, Dr. Henry Cloud pointed out that there are several building blocks for connection. The first is empathy, or the ability to enter into another person's experience in such a way that you can become aware of what they are feeling and thinking. Empathy does not mean that you are detached from your own emotions. Rather, it allows you to imagine what something must be like for the other person. Few of us will feel or respond in the same way to situations, so empathy also means that you have good internal boundaries and know it is someone else's experience, not yours.

Another component of empathy is the ability to listen and then communicate your understanding. The critical key here is conveying to the other person that we have heard them and truly understand their experience. Cloud pointed out that without letting someone know we understand what they are saying or feeling, empathy has not taken place. A simple "I hear you" or "I get that" may be all that is necessary. Also, while listening, understanding, and communicating that you understand create a connection, invalidation destroys it.

During a long-term modeling research project, one of the team members phoned the project's principal investigator and asked to discuss a potential issue involving the structure they were using to model a set of critical infrastructures. The collaborators were friends and had worked together for several years. The caller detailed a potential challenge in the way the models talked to each other, which also was related to how they presented the project to their sponsor agency. "That won't be a problem," the PI responded. "And our program manager will love it as it is." The phone call was over within a few minutes. The collaborator, although glad to have spoken up, felt like the PI thought the call wasted time and was unnecessary. For the rest of that afternoon and the remainder of the week, the collaborator felt unsettled. It would have created a sense of trust, of being heard, had the PI responded, "Thanks for voicing your concern about the model connection. While I don't think it will be a problem, I want to make sure I understand you and have us look at this together." As it was, the PI had not exhibited empathy.

SHARED INTERESTS

In the high-performance building study, the researchers found that a shared interest helped establish trust within the teams. Shared interests do not just mean that everyone wants to see a project succeed. Instead, they let the participants see how these interests benefit each person, as well as add to the greater good of the project. This type of trust allows each team member to look out for the others' interests, as well as their own. The study found that continuous interactions, coupled with a team that could see how they were bound together, were some of the most important factors in maintaining that kind of trust.

A female colleague of mine is what I would call a "connector." Just about every week, a researcher will reach out to her to see if she knows so-and-so and would be willing to connect them, because it might lead to a potential collaboration. She will respond quickly and help organize the introduction, often feeling like the linkages she establishes move science forward and make the world a better place. She will go out of her way to have people meet each other in person at conferences, often arranging for drinks and dinners to help facilitate these connections. As she says, "Nothing speeds up science collaboration more than a few drinks." This is an example of something more than shared interests. She will not be a part of the research teams themselves, but she cares about the researchers she connects.

When you are around someone who does not have your best interests at heart, you may be guarded, feeling protective of your ideas and fearful of offering anything more than that which meets the lowest bar. One example of not looking out for someone else's best interests is publishing a paper from a project without your collaborator's consent or input, even if it was decided beforehand that they did not need to be a primary author. Another example is if an agency sponsor is visiting and has an interest that overlaps with your colleague, but you don't invite them to meet. It could be argued that research is designed to be competitive, which is true. But the benefits of collaboration, which far outweigh any negatives, are crucial for innovations and problem solving in science.

In discussing the ability to trust and be trustworthy, Cloud suggested that people fell into three categories. The first kind are paranoid, exemplified by the engineer who betrayed colleagues on the team. The second type are not suspicious, nor do they expect things to go poorly. They desire trust and strong relationships, as long as they are being treated well. The third sort give trust freely, as part of who they are, because they honestly want the best for people. They seem not to fear others, unless a person demonstrates a reason why they are not to be trusted.

VULNERABILITY

Vulnerability is a hot topic nowadays, one brought to the forefront by Dr. Brené Brown, a professor and vulnerability researcher at the University of Houston. Brown's research on the dynamics of vulnerability in relationships has led to life-changing insights into its importance. There is a dichotomy here between vulnerability and power. If someone is too vulnerable, they should not be entrusted with power. But if a powerful person has too little vulnerability, we feel like they could never understand us or be empathetic. Vulnerability can be about admitting mistakes—accepting accountability for and taking ownership of them.

One day several years ago, about a week before a deliverable was due on the results of a project for a big sponsor, the research team held a meeting to go over the final actions needed. A large experiment had been run, and the data needed to be moved to the main server, so everyone on the team could access it and perform their respective analyses for the report. At the meeting, the male researcher in charge of the server told everyone that the data would be transferred that evening. By 5:00 pm the next day, nothing had happened yet. Several panicked team members called and emailed him, but got no response until the following day. When asked what had happened, he responded angrily that everything was fine, and the data transfer would be taken care of at that point. After a few weeks, with the report turned in, the team met in person. The researcher in charge of the server was in better spirits but still seemed a little down. When asked, he finally confessed that the night the data was supposed to be moved, his son became ill and had to be taken to the hospital. The boy was now better, but he was slowly recovering from a serious infection. When asked why he didn't mention the family crisis right away, the researcher said he was ashamed of not being able to deliver on a promise. The rest of the team members responded that they often felt ashamed about not working hard enough, or getting things done quickly enough, and they made a pact that nobody on the team should feel that way again.

Vulnerability is inherent in a person's need. Yet when that need is shared, more trust is gained. This can also be true at the team level, when a leader communicates to each member that they are needed for the project. If one or two people on a team use trust-building techniques, and if the entire team can then talk about trust and practice developing it themselves, then the group can potentially move closer to having a dream team collaboration.

Covey also discussed trust accounts between people and within teams. Not all deposits and withdrawals are equal. Everyone may have a different idea of what a deposit into or withdrawal from the account is. For example, you may want to highlight an accomplishment of someone on the team and celebrate their performance, but this person is shy and does not want the extra attention. Doing so anyway would be considered a withdrawal from that person's trust account. Each interaction we have with team members is an opportunity for a deposit or a withdrawal. Since withdrawals are typically bigger than deposits, stopping withdrawals may be a quick way to increase trust.

RESPECT AND CLARITY

For anyone desirous of learning more about trust in relationships, I would recommend reading Covey's *The Speed of Trust*. In it, he detailed thirteen behaviors to increase trust. For trust within research collaborations, two of them are especially important. The first is to show respect. While there is a distinct hierarchy in science, from graduate student to postdoc to associate professor or senior scientist, it should not result in disrespect or toxic behavior. One example of the latter includes being yelled at—told you are awful and that you need to "pull up your big-girl panties and deal with it." Another, according to Covey, is "being told to your face that everything is fine, but hearing from others that the same person is saying the opposite about you to peers, future collaborators, and potential employer[s]." I am sure each of us can recall examples of toxic behavior in science, and the core of this misbehavior is a lack of respect. We try to work with people who create and promote a challenging yet healthy work environment. The time has come in science for *everyone* to be able to safely call out toxic behavior and talk about mutual respect.

The second behavior relevant to research collaboration is to clarify expectations. Even the tiniest bit of attention to details, such as including a deadline in an email, can help with this process by saving time and increasing productivity. Most research project expectations center around the proposal and its statement of work. If it is important to keep the initial results of an experiment quiet until more validation can be done, that expectation should be explicitly stated, and team members should make a commitment to not share the information. In multidisciplinary work, jargon often needs to be defined to make sure everyone is on the same page. One of the quickest and most consistent ways of clarifying expectations is to keep asking who does what when, and to end each meeting with a list of the next actions to be taken. And sometimes expectations need to be renegotiated within teams. These procedures may seem elementary, but it is surprising how many miscommunications are based on ill-defined expectations. It is easy to get carried away in the technical details of a research project and forget to tell your team that the sponsors want quarterly reports, or that everyone is expected to attend a training session to be in compliance with the project's requirements.

The first known use of the phrase "being on the same page" was in 1974, but its origin is unclear. Some believe it started in schools, because it's often necessary for teachers and students look at identical pages in their workbooks. The one thing I know for sure is that having everyone be on the same page is less stressful and feels good.

Exercise

Here are some questions to ask yourself and to think about. You might want to try writing down your thoughts in a journal.

1. Can you think of an example when you did not trust a collaborator or colleague? What was it that you did not trust?
2. Can you think of an example when you were not trustworthy? What circumstances led to it?
3. What type of trust character do you think you are, and why? Are you paranoid, win-win, or all-trusting?
4. Can you think of an example where you feel vulnerable with colleagues or in a work setting?

CHAPTER 6
COMPETENCE

> Trust has two dimensions: competence and integrity. We will forgive mistakes of competence. Mistakes of integrity are harder to overcome.
>
> SIMON SINEK, *author and motivational speaker*

In the late 1950s, a series of discussions began between the United States and the Soviet Union with regard to a potential nuclear test ban treaty. Officials from both nations had come to believe that the nuclear arms race was reaching a dangerous level. Public protest against the atmospheric testing of nuclear weapons was gaining strength. Nevertheless, talks between these two nations (later joined by Great Britain) dragged on for years, usually collapsing over the issue of verification. During these discussions, it became clear that clandestine nuclear tests could be performed beyond the Earth's atmosphere, in order to avoid detection. Because of negotiations toward the treaty and concerns over the possibility of exoatmospheric testing, in 1959 Los Alamos National Laboratory (named Los Alamos Scientific Laboratory at that time) was charged with the development of a satellite-borne system to detect nuclear devices detonated in space. In those early days of space experimentation, the Los Alamos program provided planning for a total of five satellite launches, each placing a pair of satellites in orbit, in order to ensure success for each launch. The first of these launches was in 1963. Over the next seven years, the program was successful beyond all expectations, and spare hardware was assembled for a sixth launch in 1970.

The international discussions later stalled, but the Cuban missile crisis provided a major impetus for restarting the talks. In October 1962, the

Soviets attempted to install nuclear-capable missiles in Cuba, bringing the Soviet Union and the United States to the brink of a nuclear war. Cooler heads prevailed and the crisis passed, but the other possible scenarios were not lost on either US or Soviet officials. In June 1963, the test ban negotiations resumed, with compromises on all sides. On August 5, British, American, and Russian representatives signed the Partial Nuclear Test Ban Treaty. The first US satellites were also launched that year, to detect nuclear testing beyond the atmosphere. The satellites were part of Project Vela, a research program established to design methods to monitor compliance with the 1963 test ban treaty.

These satellites were intended to only detect bursts of radiation from nuclear tests, since it was, at the time, inconceivable that even nearby stars could produce detectable bursts of high-energy radiation. But Los Alamos scientists, while checking the data to show that there were not any naturally occurring bursts of radiation, discovered bursts of gamma rays coming from random directions. In the following years, scientists around the world launched satellite experiments to unravel the mystery of what was causing these gamma-ray bursts. Gamma-ray bursts (or GRBs) are now known to be explosions or other electromagnetic events that occur in distant galaxies. In 1973, the first scientific paper was published on GRBs, and this discovery took the world by surprise.

A few years later, in 1979, one of the scientists in the gamma-ray group was a young researcher at Los Alamos, Dr. Ed Fenimore. Scientists at the University of Chicago, Harvard, and the Japanese space agency now worked with Los Alamos, sharing students who traveled to this remote outpost in New Mexico, where they worked with Fenimore and the other Los Alamos scientists. Through these students, the scientists from diverse locales grew to know each and develop working relationships. Groups that normally would have been competitors instead were collaborators. A feeling of community developed through these shared students, as well as a sense that the scientists were all in this together. By the mid-1980s, the field of gamma-ray burst astrophysics was dominated by these collaborative teams. At the annual international conference on GRBs, involving 200 people, 10% of the talks were often given by members of this collaboration.

There was great competition to be the first to explain GRBs, but it took 30 years; more than a dozen satellite experiments by the Soviet Union, Europe, Japan, and the United States; plus specially designed, ground-based telescopes spread around the world. Many students who were part of this collaborative team went on to have successful scientific careers. As Fenimore described it, he felt a strong responsibility to teach the next generation, and collaborations that share students solidify the relationships between researchers. Such close relationships began with individual researchers and their students and then expanded to the other students' respective advisors. The shared students helped develop trust between the groups and built a sense of community, and the group's members realized that they could go further and faster together.

Although the shared students fostered collaborations, there still could be intense competition. In 1996, two satellites were launched to study gamma-ray bursts. The one from NASA was designed by a team from MIT, the University of Chicago, Japan, France, and Los Alamos. The other was a Netherlands-Italian team. Both used an imaging technique invented by Fenimore. The Netherlands-Italian team included a former student of Fenimore who had been mentored in this technique. As luck would have it, the American satellite failed to go into orbit correctly. The Netherlands-Italian satellite went on to make the breakthrough discovery that finally led to an accepted explanation of gamma-ray bursts. The satellite image of a GRB showed that it came from a far-away galaxy. Soon there was consensus that GRBs are the signature of the birth of black holes in some of the first stars created after the big bang. In this instance, sharing a student led to another team making the discovery of a lifetime.

With trillions of stars in every direction, there was a mere 10% chance of discovering gamma-ray bursts. What helped improve the chances of that discovery was a combination of competition and collaboration. Competition drives the urgency of a discovery or a mission. Collaboration brings together different perspectives and mindsets that provide the missing piece of the puzzle. In the 1990s, after the Netherlands-Italian satellite discovered that gamma rays were accompanied by optical bursts, gamma-ray satellites and robotic, ground-based optical telescopes were developed that automatically sent their results to whomever wanted the data. If a satellite saw a new burst, ground-based telescopes could start observing it while it was still active, which often happened within tens of seconds. But this meant that everyone had the same data. In one case, a satellite saw a burst, and follow-up ground-based observations occurred within seconds. Within a week, two papers were published explaining the burst, with the Los Alamos team submitting their paper only four hours before the CalTech team submitted theirs.

When Fenimore talks about the discovery of GRBs, he does not discuss the stars, the data, or the Vela satellites. Instead he talks about the students and the triangle of friendships that developed between the senior researchers and their students. It is this community, with everyone working toward a common goal, that Fenimore remembered.

There are different types of radiation, from radio waves, microwaves, infrared and ultraviolet radiation, x-rays, and all the way up to gamma rays. The shorter the wavelength, the higher the energy. Ultraviolet (UV) radiation is what we experience when we get a suntan. It contains more energy than visible light. This greater amount of energy can lead to sunburns and cancer. X-rays have an even shorter wavelength (between .01 and 10 nanometers), compared with UV radiation's 100–400 nanometers. The most energy-filled radiation known in nature is gamma radiation, with its wavelengths being shorter than a hundred-millionth of a millimeter. Exposure to gamma radiation is what leads to radiation sickness. Gamma radiation is rare on Earth, although uranium ore can emit small quantities of this type of radiation. The gamma ray photon is one of the byproducts of radioactive fission processes, where uranium atoms break apart to form lighter atomic nuclei. In the universe, tremendous amounts of gamma radiation are produced by solar flares and supernovas. Thankfully, the Earth's atmosphere acts as a natural shield against gamma radiation. Because of the thickness of our atmosphere, gamma rays lose their energy as they hit atoms along the way. Therefore, gamma radiation can only be observed by using detectors on high-flying balloons or satellites, such as the Vela satellites. Gamma rays travel at the speed of light, around 300,000 kilometers per second. After Vela satellites 5 and 6 went up, the different satellites could then compare the results of measuring the GRBs in their various locations in orbit, and they could finally estimate the direction the gamma rays were coming from.

"We thought it would be easy," Fenimore said, "because there were no other gamma-ray sources. But there were a lot of bumps in the data, showing up on multiple satellites. It was 1973 before it was realized that these were real objects, gamma-ray bursts, out there. It was such a surprise because of the feeling that stars are constant. In the beginning we couldn't figure out how any star produced gamma rays. Soon there were two ideas: one that they were nearby, in our galaxy, and the other that they were very remote and much more powerful. The consensus was that they were in our galaxy."

One particularly successful satellite was the Swift satellite, which used an array of 52,000 pinholes that produced thousands of overlapping images. Mathematical algorithms decoded this complexity and revealed the source of each burst. The next step, after 1973, involved the construction of robotic telescopes at locations around the world, including one in the Jemez Mountains of New Mexico. There were also installations all around the equator, so the satellites would always be "visible." Fenimore said GRBs occur about once a day somewhere in the universe. On October 9, 2022, the brightest GRB ever was detected, bright enough that smaller amateur telescopes could spot its visible-light emission. Astronomers quickly determined its distance and found that it was the closest such burst yet seen. It was located a mere 2 billion light-years from Earth.

In science, there are different types and levels of competence, with the research on GRBs being a higher-end example. When most of us think of this term, we envision tests designed to measure one's intelligence and skills after academic courses are completed, or to measure one's overall scholarly ability. Mathematical competence, as defined in the recommendation of the European Parliament and the Council of 18 December 2006, is "the ability to develop and apply mathematical thinking in order to solve a range of problems in everyday situations. Building on a sound mastery of numeracy, the emphasis is on process and activity, as well as knowledge. Mathematical competence involves, to different degrees, the ability and willingness to use mathematical modes of thought (logical and spatial thinking) and presentation (formulas, models, constructs, graphs, charts)." This document goes on to define competence in science as "the ability and willingness to use the body of knowledge and methodology employed to explain the natural world, in order to identify questions and to draw evidence-based conclusions."

While there may be many types and levels of competencies in science, it is clearly one of the most important indicators of a researcher's success. We can think of it as mathematical ability, computer coding skills, an understanding of key theories in a field, technological competence, and writing skills, or the ability to present and communicate ideas and results. Another word for "competence" is "capability." The teams from around the world trying to unravel the mystery of GRBs had extremely high levels of both scientific competence and capabilities: building satellites and robotic ground-based telescopes, performing data analyses, and constructing physics models, to name just a few. Even then, it took 30 years before a consensus was achieved.

Both character and competence go into helping to build trust, as discussed in the previous chapter. Character means your integrity, your motives, and your intentions toward others. Competence involves your skills, your results, and your overall track record. While this chapter is focused on the two competence cores—capabilities and results—in the trust stool, all four cores are critical for building and maintaining trust in teams. We've all been aware of researchers with enormous capability, strong results, and an insanely high Hirsch index. Unfortunately, their "ends justify the means" mentality means that their projects end up cutting corners, and they sometimes may work in dishonest or unprincipled ways. On the other hand, to only have integrity, and not the other three core features of trust, is to be a "nice person" who is basically useless.

Using Covey's tree metaphor, capabilities are the branches that produce the fruits, or the results. Capabilities are essential in science. Although it is important to know the basic calculations for statistical tests, fewer students now have to write out and solve equations on paper. Their capabilities are moving more and more toward computer science. Technology and the globalization of research are outdating skill sets faster than ever before. Mid- and late-career researchers are keenly aware of what it takes to keep up in their fields: new technologies that can be applied to their work, and ever-changing computer applications and codes for analyses and statistics.

There are several types of capabilities when it comes to competence: talents, skills, knowledge, and experience. Talents are our natural gifts and strengths. Skills are our proficiencies, the things we can do well. Knowledge and experience represent our learning, insight, understanding, and awareness. Style depicts our approach and our personality. It indicates how we bring together our capabilities and align our natural gifts, passions, skills, and knowledge. Style gives us an opportunity to learn, to contribute, and to make a difference. Results are represented by our performance, credibility, and reputation. This is why an annual review is often called a performance review. What results did you get at work? How well did you perform? In science, the first part of our career—in college, graduate schools, and postdocs—is spent enlarging our expertise and capabilities. The next part is devoted to maintaining that expertise and keeping it relevant. Trying to keep up with the fast pace of scientific outputs, technologies, and developments in computing, all of which are part of the scientific revolution, can be overwhelming.

Results matter immensely in science. They add to your credibility and influence whether people view you as competent. The importance of results is what has led to the paradigm of "publish or perish." The results section is the core of all scientific publications. We work hard every day to obtain results in the laboratory, the field, and our data analyses. Results are the fruits on the tree—the tangible, measurable product of the roots, trunk, and branches. Sometimes we have negative results in our experiments, leading to a rejection of our null hypothesis, which may be a more important outcome than a positive result. Without results, it is not research.

Results also cover other aspects of research that fall outside of a project's end result, which are known as follow-through. Did you send the promised data files to your colleague this week? Do you submit your monthly or quarterly reports to your research sponsors on time? Are the aspects of your research compliant with all permits, approvals, and other safety considerations?

There are three key indicators in evaluating results: past performance, present performance, and anticipated future performance. For a collaborative project, team members will look to past performance to support their faith that you will do what you say you will. It breaks down trust over time if, in a team meeting, one person often states that they can add to the introductory section of a paper or perform an additional statistical analysis on the data, but then never gets around to it without being hounded by the other members.

I once had an experience on a modeling team where the leader signed us up for a week-long, off-site retreat. During our time together, we went through many exercises, and it was much like team therapy. While the focus was on discovering your "true north" and becoming an authentic leader, participating in a retreat as a team was a new experience for all of us. This was not something that most scientists at a national physics-based laboratory do together. I give credit to the principal investigator of the project for having the courage to take us away from the actual work at hand. One of the most important things we learned on this retreat was the idea of responsible commitment. For everything you agree to do for a project or for the team, you should ask yourself, "Can I really do this within my stated timeframe?" In other words, "Can I responsibly commit to do what I say I can and will do?" After that retreat, the team members often not only found themselves doing this, but they also asked each other, if they committed to doing something, "Are you sure you can responsibly commit to this?" Competence is doing what you say you will in the timeframe within which you state you will do it.

If you cannot responsibly commit to doing something, then having an open dialog with the project team is critical for solving the problem. Perhaps the analysis is outside of your expertise in R computer code, and you need a few extra days to learn how to do it. Or maybe you are overwhelmed with other things at the moment and would not have the time required to complete the task. All of us don't want to let down our colleagues, and that is what induces us to open our mouths and offer to help. The next time you have that inclination, before you start to speak, ask yourself the question, "Can I responsibly commit to this?"

Given the importance of results in establishing credibility and trust, both in ourselves and with others, another crucial question is, "How can we improve our results?" One clear way is to not commit to doing things that you feel you might not be able to get around to doing. Scientists are overwhelmed with work, as well as opportunities, and most of us have experience in saying no to things. Still, learning to get better at saying no to distractions, and focusing instead on the tasks that are part of a research collaboration, will help bring your attention back to the actual science. That seems easy, but in science the pressures are intense: to keep finding funding, publish results from projects that have ended, review papers, be an active member of a professional society, serve on committees, and do all the other things researchers are expected to do. It is, in fact, overwhelming. In each of these areas, asking yourself if you can responsibly commit to a task can be lifesaving, and the end result will be trust in your word, both for yourself and your team members.

Exercise

Here are some questions to ask yourself and to think about. You might want to try writing down your thoughts in a journal.

1. Can you remember a time when you did not follow through on a commitment? What were the reasons for not doing so?
2. What is an area of growth you are continuing to learn about, either in your own field of science or in another one? How do you keep up with technical advances in your field?
3. Have you had experience with questioning a colleague's competence? What were the things that made you feel that way and led you to question their abilities?

CHAPTER 7
COMMUNICATION

> The single biggest problem with communication is the illusion that it has taken place.
>
> GEORGE BERNARD SHAW

I received a phone call one day, on the weekend and at my home number, from the other PI on a collaborative project. I remember hanging up the phone and thinking the conversation was a bit odd. It was about a potential conflict in our collaborative group. In this case, we were the two PIs, coming from larger institutions. The potential conflict was between the two postdocs on the project. Both sides of the project team had an outstanding postdoc, each with their own personality, technical expertise, and ideas for the goals of the project. As the PI for my half, I had not really seen any conflict between these two individuals, other than a general discussion on how the project should go about answering the proposed hypotheses.

The other PI had called to discuss a manuscript that the team had completed, concerning a large experiment we had all done together. This experiment involved both field sampling and laboratory work to answer a few specific hypotheses. There were rigorous and specific protocols for sample collection, to maintain the quality of both the samples and data. The two postdocs, plus myself and another colleague, set up the study and completed the sampling work together at my institution. We completed that work over several days. The samples were then sent back to the collaborating institution with the other postdoc, to be analyzed at their laboratory. The two postdocs subsequently worked on the resulting manuscript together, with the other institution taking the lead.

The first manuscript from the project was a methods paper, as there were some novel techniques that were employed with the samples. The manuscript was sent to a journal and was rejected, but it came back with some good comments from reviewers on how to improve the paper. Our colleagues from the other institution had communicated to us that they were going to wait many months, or maybe even a year, to get back to the manuscript, as they needed to focus on other work.

Shortly after this I received that weekend phone call from the other PI, wanting to talk about the manuscript. They had been working on it and had incorporated the reviewers' comments on improving the paper. They had even made a cool little video of the results, to go along with the text. It sounded like great news to me. The main purpose of the call, however, was to talk about authorship. There was concern from the other postdoc that our postdoc should not be an author on the paper. There was also some conflict between the two postdocs. Not only did this surprise me, but I was also astonished that it did surprise me. I felt like I had great communications with my postdoc and a fairly good relationship with the other postdoc. I am usually acutely aware of conflict. Since I don't prefer it, I am also super aware of potential conflict on the horizon.

I responded that I felt my postdoc deserved to be on the paper as an author. When I think about authorship, I ask the question, "Would this paper be possible and publishable in its current format without this person?" If the answer is yes, then that person could possibly be acknowledged but not be on the author list. This method is not foolproof, but it can help decide authorship. The answer to this question for my postdoc was a huge and absolute NO; this paper would not have been possible without that person's efforts. I pointed out all the hard work my postdoc had done on the study and painted an optimistic picture of how the team could easily work through what came down to personality differences. I concluded the conversation by stating that my postdoc "deserved to be on the paper even more than I did." This, to me, made an obvious case for that individual's inclusion, as I had done weeks and weeks of work, as had the other PI, to support this experiment and paper. I thought the conversation had gone well, and I felt good as I hung up the phone in my office at home.

Four or five weeks later, I got an email from my collaborator with the subject line "Great news"—always a terrific way to make someone eager to open an email. It said that our manuscript had been accepted by a good journal. It was now published online. "Wow, exciting," I thought. I clicked on the link and opened the published paper. At the same time, though, I had the thought, "Huh, I did not know they had sent this to a new journal." Or that it had been accepted. I quickly saw my postdoc's name on the author list and breathed a sigh of relief. I saw everyone's name on the paper as an author. Except there was one missing—mine. A PI for the study. The person who originally saw the grant opportunity, developed a strategy for a collaboration with my postdoc, wrote most of the proposal with my postdoc, and participated in all of the work that took place at our institution. Nonetheless, I was merely listed in the acknowledgments.

Instead of responding right away in an email—which would have just been a three-word question, "WTF?"—I waited a day and then wrote back to the postdoc at the collaborating institution, with a cc to the other PI. I included the entire list of over 30 major things I had done, both for the research idea and the work that eventually led to the manuscript. The PI immediately responded, explaining that my omission as an author was not the postdoc's fault, who took to heart the comment that my postdoc deserved to be on the paper even more than myself. "Oh," I thought, "that person missed that I was being facetious." We resolved to talk on the phone again to rectify the problem, but we never did. The mix-up occurred at the end of the grant's performance period, and the damage had already been done.

I share this story because it is a great example of a miscommunication, as well as a lack of overall communication. While it was a painful experience for me to go through at the time, I also wanted to use it as a learning opportunity. Even if a collaborator is not highly communicative to begin with, what could I have done differently to make sure this did not happen? How could I have responded better, as well as in a more positive way? How and why did I assume the phone call resolved all the issues?

Misunderstood or misidentified expectations are one of the most common issues that come up in team projects. While some communication tools are useful in one-on-one situations, in larger teams it can be more difficult for everyone to clearly understand their specific roles and the expectations of the project. As we increase the number of people working together, every individual brings their own expectations to the table. In a one-on-one arrangement, a person can learn to repeat back what they have heard to the other participant, inquire if they heard things correctly, and ask additional questions. While this can also be done in a group setting, there may be conflicting expectations within the team that may differ from those of the principal investigator. Or maybe several team members missed a meeting where the expectations were clarified.

Another common communication challenge is the need for an expectations pivot. As unanticipated results from experiments come in, or challenges develop with field or laboratory work, the research objectives must be modified to meet the new situation. Expectation pivots are especially challenging, for several reasons. People on the team can be confused by them, particularly if they are not always communicated clearly. Various collaborators can have different ideas about how to best execute the pivot. A great example of this was the COVID-19 pandemic, where research teams worldwide struggled to keep going during extreme circumstances in which laboratories and field teams were shut down and meetings became virtual. Supply chain issues and inflation in 2022 and beyond also impacted research projects around the globe. Yet deliverables were still due, and the ambitious timelines of projects were impacted.

One solution is to convey the expectations of a project clearly—and as often as possible. It may seem inane to communicate them more than once, but this is one of the common pitfalls for team projects. I once had an academic researcher tell me about a US Department of Defense project to develop a "green" emergency flare for the sides of roads. The professor's team created technologies that reduced the use of chemicals and toxicants, as well as the carbon footprint of the products, and developed an emergency flare that indeed was more environmentally friendly. When the team demonstrated their successful end product to the sponsor, one of the federal participants asked why the flare did not have a green flame. After pulling out the statement of work, it could clearly be seen that the team was asked to "create a green flare." This may be one of the more amusing examples of misunderstood expectations, but small miscommunications happen every day in projects. Over time they can add up.

The best time to clarify expectations is at the beginning of a project, although it is never too late for a team to ask questions about them. Sticking to clear initial expectations will depend on the project and the situation, but having a team that is up-to-date with potential solutions to a problem that has certain constraints may help speed up any necessary pivots. One of the goals in having a group discussion about expectations is to identify any potential challenges and roadblocks up front. There are several questions that can be asked periodically to gauge where everyone is with regard to expectations, as well as to identify tiny roadblocks that can make the team move slower and be less productive. Communication with the honest intent of helping the team succeed and clarifying expectations is like adding super-high-octane fuel to a sports car.

One of the best ways to communicate effectively is in person. Other communication tools lose some of their effectiveness the farther away they get from being delivered in person. As social animals, we have evolved to read other people's expressions, gestures, and any other small thing that could add nuances to a person's true intentions. When we're face to face, if we joke with someone, we can smile and give other indicators that we are teasing. In emails, we have reverted to using the smiley emoji to help soften things and communicate the subtle clues we would otherwise be expressing in person. Just like group brainstorming sessions, an expectations check-in with a team is best done in person or in a videoconference. Having a session about group expectations and project aims that is facilitated by one of the leaders can also help elicit feedback from the quieter introverts on a team. With research projects, a meeting to set expectations can easily become a creative brainstorming session. And, like everything in science, it is always improved if a whiteboard is present.

Over the past several years I have created a series of questions that could help a research team see where they are having difficulties in all team members understanding a project's expectations. The exercise for what I've called project expectations and aims review, or PEAR, consists of a series of questions that are best asked at the beginning of a project. They can also be repeated at any time during the life of a project, and it would be good to review the questions, as a team, either quarterly or annually. Once a team is used to communicating expectations, its members become more trusting and are open to asking questions when something is not clear. While most scientists in the last 30 years have been taught that is okay to say "I don't know," it is still challenging for many researchers to openly show a lack of knowledge. The PEAR exercise can also help build trust within a team by encouraging open communication.

One of my most recent experiences with leading a PEAR exercise was with my own team. It involved a sizeable project, with over 25 team members. We were tackling the challenge of merging huge existing climate models with epidemiological models, in order to make short- and long-term forecasts for infectious disease outbreaks that could occur in response to expected large-scale environmental changes. The team had worked on this proposal for four years before finally getting it funded, and they were excited to begin the work. It was clear in the PEAR session, however, that nobody really understood the expectations for each of them in the project. As science teams have grown in size over time, become more diverse, and tackled more complex problems, it has become easier for individuals on a team to not quite understand their specific role within the group. Teams may not be fully in sync with the project's mission, or fail to see how all the pieces fit together. The above experience with PEAR took place during the pandemic, so the questions were sent out to everyone beforehand to answer on their own. When we discussed the questions as team, it was apparent that some of the power of answering the questions as a group, using one's gut instincts, was lost. Feeling like you don't understand what is expected of you is an isolating experience. On the other hand, seeing that others on the team may also not have clarity, and then working through that together, is empowering.

Overall, there are 15 PEAR questions, but 5 of them are the most important. These deal not only with expectations, but also with project-specific items regarding communication styles and project management. They are designed to be idea generators, and they can be modified to suit different types of projects and their individual team members.

Below are the five core PEAR questions that can help jumpstart a project or surmount a challenge the team may be facing. (All fifteen of the questions appear at the end of this chapter.)

1. Do you feel clear about your specific roles and the expectations for your particular part of the project?
2. What is unclear in any of the project's expectations for you?
3. What do you think will be the most challenging part of this research project?
4. What are some good ways of dealing with potential challenges?
5. What do you see as some follow-up actions for the team and for yourself?

An expectation is a strong belief that something will happen now or is likely to occur in the future. In 2014, scientists from the University College of London reported on a project involving 18,000 people worldwide that was designed to study what determines happiness. The investigation looked at the relationship between happiness and reward, as well as the neural processes that lead to feelings such as happiness. The authors found that moment-to-moment happiness reflects not just how well things are going, but whether things are going better than expected. The lower the expectations, the happier you may feel about something. In terms of research teams, transformative science is based on expanding expectations, and regularly taking the pulse of everyone's expectations can potentially lead to transformative results, as well as more happiness on the team.

PEAR's objective is to highlight and communicate the expectations of everyone involved in a project. This includes fully exposing the challenges that lie ahead for the team, thus allowing potential solutions to surface ahead of time. Its intention is to push a research team into becoming an exceptional team, with unlimited potential, that is more creative, productive, and better at problem solving. The focus of PEAR is to build trust and cohesiveness within the team.

Exercise

Here are the full set of questions to review about project expectations and aims.

1. What are you most excited about regarding this project?
2. What do you think the team already does well?
3. Do you feel clear about your specific roles and the expectations for your particular part of the project?
4. What is unclear in any of the project's expectations for you?
5. What do you think will be the most challenging part of this research project?
6. What are some good ways of dealing with challenges?
7. What do you see as some follow-up actions for the team and for yourself?

8. What meeting frequencies and types of meeting (large or a subgroup) do you prefer?
9. Do you have a particular software preference (Google Docs, Teamwork, or others)?
10. What has been a lesson you learned from working on a past collaboration or team?
11. What questions do you have for the team to help clarify the expectations for you and for the project?
12. Do you see any technical gaps in the team?
13. How do you like to communicate best? By phone? In person? (Never?)
14. If you had a chance to begin this project all over again, would you do things differently?
15. What does success look like for the team?

CHAPTER 8
FISH DON'T KNOW THEY'RE IN WATER

One fish asks another fish, "How's the water?" The other fish replies, "What the hell is water?"

DAVID FOSTER WALLACE, *author*

Not collaborating is always easier. But the benefit is always bigger than the costs.

BILLY KARESH, *wildlife conservationist*

When the collapse of the Soviet Union seemed imminent, the United States' administration now had to think about the risk of thousands of nuclear scientists, secret cities and laboratories, weapons-production facilities, and vast armament complexes spread across the region. Senators Sam Nunn, a Democrat from Georgia, and Richard Lugar, a Republican from Indiana, addressed this challenge by sponsoring the Soviet Nuclear Threat Reduction Act of 1991, also known as the Nunn-Lugar Act. It was signed by President George H. W. Bush two weeks before the dissolution of the USSR in December 1991.

After the fall of the former Soviet Union, 15 countries were separated out. Now there was a new challenge at hand. Not only did each country have to develop the governments and infrastructures to stand on their own, but they also had to deal with legacy issues, such as having arsenals of weapons of all types—chemical, biological, and nuclear. These weapons were unsecured and scattered all over the former Soviet states. They were inherited problems that impacted the whole globe. What the world needed was a stabilized reduction in such armaments, particularly weapons of mass destruction, to reduce the risks of war.

There were four successor states that inherited the USSR's nuclear weapons and forces—Russia, Belarus, Kazakhstan, and Ukraine—and the Nunn-Lugar Act established a US government assistance program for these newly independent states. The Cooperative Threat Reduction (CTR) program was designed to help combat the threat of the largest nonproliferation problem in the modern atomic age. Its $400 billion in annual assistance went toward safe transportation of the nuclear and other weapons to secure sites, with an inventory of and either safe storage or destruction of them. Each aspect of the program was based on cooperation and collaboration. Successful implementation of the program required coordination between US agencies, as well as with the partner countries' governments and the scientists supporting the technical mission. Little did everyone involved know at the time, but they were creating one of the most successful and rewarding collaborative science efforts of the twenty-first century.

In a rescoping review in 2005, two new programs were created within CTR. The first was designed to assist Eurasian and Central Asian nations in preventing the illicit trafficking of weapons, through the use of radiation detectors at border crossings. The second addition to the portfolio was the Biological Weapons Proliferation Prevention program. This new initiative worked primarily with Uzbekistan, Kazakhstan, Georgia, and Ukraine to strengthen their laboratory biosafety and biosecurity capacities, as well as to secure the former Soviet biological weapons facilities in these countries.

The new focus on biological weapons, or select agents, came partly from the anthrax attacks in the United States that killed 5 people and injured 17 in 2001. It also arose from knowledge of around 40 facilities that were associated with the former Soviet bioweapons program. These toxic biological samples, which were unsecured and dispersed across several countries, increased the global risk of a biological leakage or, worse, an attack. At its height, an estimated 30,000 to 40,000 scientists participated in biological weapons efforts. In 2002, the CTR program starting working with Russia to disband and decontaminate such laboratories and buildings. Its initial years included some of the most noteworthy breakthroughs in threat reductions in recent history. The successes included disassembling a dual-use laboratory for viral animal microbes in Biokombinat, Georgia; dismantling a large facility for anthrax production in Stepnogorsk, Kazakhstan; and destroying 165 tons of anthrax that were abandoned on Uzbekistan's Vozrozhdeniya Island in the Aral Sea.

Another initiative of the CTR program was to develop and support collaborative research among scientists in the United States and its partner countries. This not only helped strengthen biosafety and biosecurity capabilities within these countries' laboratories, but also brought them new technologies for the biosurveillance of pathogens of global concern. Moreover, this collaborative research could be designed to ask important questions to increase our knowledge about the ecology of infectious zoonotic disease systems. If you can understand the workings of a natural disease system in animals, humans, and the environment, then you can better detect outbreaks or find and plug the source to prevent a disease from propagating. It was win-win for all partners and scientists, as well as the world. When it comes to infectious diseases, a threat *anywhere* is a threat *everywhere*. Other countries, such as Canada, Germany, and the United Kingdom, also support their own cooperative threat reduction programs. To this day, these programs help create a safer world from infectious diseases. The years of dedication and support by participants in these programs undoubtedly helped minimize the impacts of COVID-19 in partner countries around the world, due to their development of better diagnostic capabilities and biosurveillance infrastructures. Over the years, these collaborations have led not only to cutting-edge science for detecting infectious diseases, but also sustainable collaborations, which have to be built on trust in order to survive.

WHY HELP ONE ANOTHER?

Not every international travel experience can feel like the world is watching out for you, because 99% of the time it isn't. Most of us who have traveled internationally can remember a moment when we did not have a single clue about what we were doing, because everything was so new and foreign. Imagine stepping off a plane in a country where you do not speak the language, and the new job awaiting you involves learning a technical field in science. At the customs checkpoint, you follow the mass of people and act like you know what you are doing, but everything you hear is gibberish. None of your fellow travelers speak your language, so it is no use talking to them. One such instance occurred when I was leaving Nepal. There were two customs areas, one for men and one for women, with signs in Nepalese above each. The latter area was empty, and I thought the signs meant "open" and "closed," believing I might have picked up a little Nepalese. After I stood behind the hundreds of waiting men, one of them finally waved for me to go to the empty area. Soon the rest of the men started making the same motion. When I finally realized my mistake, all the men cheered for me and jumped up and down. At first I felt triumphant. Next I felt like an idiot, like when I forget my glasses and cannot read a restaurant menu without literally putting it next to my left eyeball. Then a sense of overwhelming vulnerability came over me, and I wondered, "What I am doing wandering the world by myself?"

Many young researchers head off every year to graduate school in places where English or the country's native language is not a language they know. I cannot imagine having to master the specifics and theories of my field of study, plus scientific and statistical methods to answer research questions, while also learning a different country's language and fitting into its culture. That scenario combines all of my recurrent bad dreams about not being prepared for a test, or being naked and not being able to find my gate, because everything is written in chicken scratches and everyone is walking around clucking.

E. O. Wilson once gave an interview where he said science was like a magic barrel, where you could pull out a question to answer. If you do your job right, you get to go back to the barrel and pull out another question. Collaboration is similar. If your collaboration goes well, then you can either continue it or move on to a new one, with the reputation of being a good collaborator. One of the best examples of this is Dr. Jason Blackburn, a professor at the University of Florida who studies zoonotic pathogens, such as anthrax and brucellosis, in wildlife. He works with partners in countries around the world to strengthen those nations' biosurveillance capabilities and conduct research in host-pathogen systems, in order to make the world safer. He is considered a super collaborator, because his work connects researchers in multiple projects, institutions, and countries. He is driven to understand host-pathogen systems and is deeply interested in finding out how infectious diseases emerge and are propagated in animal populations. His enthusiasm is as infectious as the diseases he studies. His energy is unbounded, and his smile is ever present. He frequently reaches into the magic barrel of science for new questions, and he builds his family of collaborators just as often.

The projects Blackburn works on are supported by cooperative programs designed to reduce the threat of infectious diseases. He has spent years working on select agents, collaborating with researchers in former Soviet laboratories now located in countries such as Ukraine and Kazakhstan. He has labored alongside researchers in some of the most challenging and austere environments on Earth. He is not a virus hunter. He is a scientist who believes in understanding ecological systems, and in doing so collaboratively. A perfect day for Blackburn might mean digging up a feral hog that was buried by a farmer in Vietnam after its unexplained death, in order to obtain samples from it. Then, in the afternoon, working side-by-side in the laboratory with other researchers, and ending the evening with a call to an international team to work on manuscript revisions.

When I asked Blackburn how he builds such strong trust relationships so fast, he responded that he has developed a three-way process. First, he immediately and transparently shares data with his new colleagues. This could involve not only supplying actual data that might be useful in their research, but also procedures and laboratory protocols, publications, or anything else that could be helpful. He does this to break the ice and start building trust with his collaborators. Second, when he is training researcher partners in the use of a new method or type of analysis, he might do something once, but then he lets the others take over from there, so they can immediately become proficient in that skill. This also demonstrates that he trusts their competence and expertise. Third, he leads with empathy and learns about personal aspects of his collaborators. Each of these behaviors is a drop in the bucket of trust.

Becoming friends with foreign partners is one of the reasons many people work internationally. The benefits of understanding and engaging with another culture are limitless, and such actions are worthwhile in almost every context. This may seem intuitive, but most scientists who work internationally can share horror stories of fellow scientists being judgmental, noninclusive, and downright rude to their foreign partners. The term for this is "helicopter research," where scientists from wealthy nations visit low-income countries. They collect samples and publish the results with little or no involvement from local researchers. There have been recent calls to end this practice and to have more meaningful collaborations that have positive impacts for the local country.

DIPLOMACY AND EMPATHY

An aspect of collaborating that Blackburn best understands is the power of diplomacy. One of its definitions is the art of dealing with people in a sensitive and effective way. Blackburn understands that diplomacy means playing the long game. It is an art that can be practiced and learned. You may ask yourself, "What is an example of diplomacy in my daily collaborations?" It could be not reprimanding a team member while you both are around others. It might mean trying to better understand why a team member needs to take personal time off, or why they missed doing something they had previously committed to. Or perhaps exhibiting inclusive behavior with everyone, regardless of any disparities. Diplomacy is the difference between sensitively addressing something and bluntly calling it out. It means maintaining trust when tackling disputes and difficult conversations. It involves having and showing empathy toward others. And it can also allow those treated diplomatically to complete exceptional work. Blackburn has mastered the behavior of having empathy for others, yet standing up for himself and his values or beliefs, as well as those of the team, when needed.

Empathy is another powerful tool when working with any collaborators, including those who come from different cultures and backgrounds. For the most part, empathy refers to the various ways we respond to each other. One facet is putting yourself in someone else's shoes, which can lead to a greater understanding and appreciation of the challenges they have overcome. Over the years, there has been a slow decline of empathy in our society. As we become more engrossed with our hand-held technologies, and more people now live alone, our empathy muscles are getting weaker. Then we watch the news. I have yet to meet anyone who, after doing so, exclaims, "I feel so much better about the world, my fellow humans, and myself." All of this can lead to compassion fatigue, which is the feeling that you have no more empathy left to give.

The best analogy I have ever seen about the state of empathy comes from Dr. Jamil Zaki, in *The War for Kindness: Building Empathy in a Fractured World*: "Being a psychologist studying empathy today is like a climatologist studying polar ice. Each year we discover more about how valuable it is, just as it is receding all around us." Zaki is a professor of psychology at Stanford University and the director of the Stanford Social Neuroscience Lab. He and his colleagues have developed methods for measuring empathy and determining how people can learn to empathize more effectively. Among other examples, this book included the noteworthy work of Sara Konrath's team, which found that the average American in 2009 was less empathic than 75% of Americans just 30 years earlier. Zaki created a set of kindness challenges, which are exercises at an "empathy gym" that can strengthen your empathy muscles. By making use of them, researchers can help create a kinder world in science.

Interactions between humans are both verbal and nonverbal, with the latter offering subtle clues through body language, facial expressions, and voice volume. We can often tell when our colleagues are tired, feeling down, or distracted. In our face-to-face interactions, we gather information about both the other person and their attitudes toward us. This means that we constantly ask ourselves, "What does the other person's response say about me?" Small gestures, such as a touch on the arm or shoulder, can comfort us, and a smile can literally be worth a million dollars. What these gestures and signals can communicate is empathy. They show that we hear what the other person is saying, and we care.

There are three primary types of empathy. The first—identifying what another person feels—is called cognitive empathy. The second, known as emotional empathy, is sharing our emotions with others. The third is wishing to improve the experiences of others, which is called empathic concern. In emotional empathy, we often take on the feelings we are observing in others. When we see a friend or colleague who is upset, we begin asking ourselves questions about how are they feeling. Are they really upset, or just tired? What will their next response be? In other words, we are cognitively trying to figure out what is affecting our friend. With empathic concern, we may now launch into action by offering suggestions to help our friend get through the challenge. Or we take our friend out to get a breakfast burrito, which seems to cure any ailment.

While it can sometimes be helpful to take on someone's pain briefly, it may not be that useful to climb down this dark hole with a friend. Unless you bring a flashlight and some snacks, which would demonstrate empathic concern. Instead, reach out a hand and validate that person's feelings by saying, "I see you have fallen and that must be very scary. Here is my hand, and let's see if we can get you out. Lassie is here to help pull on the rope." On a research project, this might mean asking how a colleague or a student is doing and then listening to the reply. But first we must remember to look up from our laptop and data and see that we are working alongside other humans. Even with the most annoying and aggravating colleagues, asking ourselves why are they so angry all the time can lead us to inquire, "How are you doing?" The way in which they respond can give us a better understanding of their behavior, which can soften our responses to them.

Several of my colleagues have pointed out to me that traits like empathy and kindness are fixed. They mention knowing coworkers who have zero empathy and would not lift a hand to help save a kitten, let alone a graduate student or a collaborator. Such individuals seem indifferent to other people, and their general demeanor is one of righteousness and contempt. While many behavioral traits can be thought of as fixed, it is important to realize that there are ranges around that fixed base. These ranges are affected by our experiences, relationships, behaviors, and the ideas we are exposed to through media. I may be a mesomorph, fixated on a high-protein diet, but my eating and exercise behaviors can lead me to a thin or overweight mesomorph. The same is true of IQ, in that it can be raised over time with healthy habits and through education. Thanks to Carol Dweck's studies on mindsets, covered in her 2017 book, we can now realize that how we see ourselves and the world can affect everything. If we believe that traits are set in stone and we are stuck with our basic intelligence or looks, then that is a fixed mindset. If we believe we get good grades because we worked hard, not because we are smart, then that is a growth mindset. One of the most important findings from Carol Dweck and her colleagues was that our mindset can be changed.

Empathy may be fixed, but it is also a malleable trait that can be grown and nurtured, or shrunken through exposure to a lack of empathy by others. In one 2014 study by Drs. Karina Schumann, Jamil Zaki, and Carol Dweck, the team wanted to test if people who believed empathy was a skill that could be cultivated might try to foster empathy in themselves and learn how to empathize with others. In the study, they asked hundreds of people which statements they agreed with most:

In general, someone can change how empathetic a person they are.

In general, someone cannot change how empathetic a person they are.

They reported that the participants were divided about 50/50 regarding which statement they most identified with.

When the participants were put through a series of situations used to measure empathic responses, the team found that people with a growth mindset worked harder to try and empathize with others, including those who were different from themselves. The team was also able to change the empathy mindset of the participants by showing them stories about someone's empathy that depicted either a fixed or a growth mindset. The study participants believed each story. After reading the growth story, the fixed-mindset group gravitated toward empathizing more with others. In a short time, the team had pushed the participants to the upper or the lower end of their empathetic range. This research has shown us that traits such as empathy, kindness, and compassion are up to us. We can decide what side of the range around that middle point of empathy we want to be on.

It is a common belief that women are more empathetic than men, and some studies have shown that to be true. One example is a 2001 investigation where the researchers showed study volunteers a video of people telling emotional stories. They asked the participants to guess how the speakers felt, and women indeed did guess more accurately than men. In the next part of the study, the researchers told the participants they would be paid money for each answer they got correct, thus shifting their motivation. This slight change erased all differences between the men and women, showing that motivation through incentives can encourage people to become more empathetic. If we focus on wanting to be empathetic, we can be.

Effective collaboration is fueled not only by communication, but also by empathy. Knowing that empathy can increase trust between collaborators and improve performance might be a nudge that could increase empathy in research groups. In one study by Adam Grant, his team found that empathetic interventions could dramatically increase productivity, since the feeling that one's work benefits others is closely related to empathy. In science, empathy may be a clear driver for researchers working on global or local challenges, thus trying to make the world a better place. Empathy also affects how we impact the people around us, including those on our teams and in our organizations. Perceived social worth is the degree to which employees feel that their contributions are valued by other people. Perceived social impact is more about how employees believe their actions benefit others. Grant's study provided evidence for the idea that job performance can be enhanced by perceptions of task significance, which are judgments that one's job has a positive impact on other people. He also showed the value of stories and narratives as corrective lenses for reframing our experiences.

IT'S NOT WHAT YOU DO, BUT HOW YOU DO IT

In the past few decades, the world has become hyperconnected and hyper-transparent. This new reality has led to, and been a part of, the scientific revolution discussed earlier. Information is shared instantly and potentially globally. Scientific talks can now be discussed in real time via Twitter, or gossip can be shared across the table in a meeting through a text message. The main argument in Dov Seidman's book *How: Why How We Do Anything Means Everything* parallels the words of Hall of Fame basketball coach John Wooden: "It's not what you do, but how you do it." For example, how we ask a colleague a question can be crucial in maintaining trust and the integrity of the relationship. If we use the words "Did you get that data to me yet?" versus "Have you had a chance to send the data?" we are asking the same question, but the phrasing can elicit different responses, one more defensive than the other. On the flip side, how we answer questions can also be important. We all know someone who replies to every question with a defensive answer, accompanied by either annoyed, superior, hurt, or offended body language. Everyone quickly learns to not ask that person questions, which slows down team cohesiveness.

While most science projects will have a principal investigator or team leader, all members of a team are leaders, both through their own responses to coworkers and to challenges in a project. As self-governing individuals, we can approach everything we do from a leadership perspective, whether it involves walking into a meeting and being engaged, or writing an article or an email. Every day, we can consciously align our behaviors with our values. Feeling like a leader can also give us confidence. We are all role models for those around us, whether they are senior scientists or the most junior undergraduate students. Leadership is not a title on a business card. Rather, it is our mindset and the way in which we approach the situations and relationships that make up our days. Leadership is thinking about the "how" and upholding our personal values. It means changing mindsets, so everyone on a team feels like a leader, and the more-senior team members see the younger or less experienced participants as leaders.

INTERNATIONAL COLLABORATION

When I first posed the question, "What makes for a successful scientific collaboration?" to a small group of scientists in Tbilisi, Georgia, they looked like deer in the headlights. That is not to say that they did not have phenomenal answers to my question. But it took them a little while to wrap their head around the different facets of working cooperatively with other researchers. For these Georgians, collaboration is integral to who they are. It is also central to their careers, where they engage with others in countries around the world to reduce the threat of infectious diseases in humans and animals. For them, a successful scientific collaboration regarding infectious disease outbreaks is about understanding that we are all in this together. Asking them about collaboration was like asking a fish, "How's the water?" I should have known this from when I first flew into Tbilisi and saw the sign that greets visitors: "Tbilisi, the City That Loves You."

Successful international cooperation stands on a foundation of trust, and while that is somewhat easier between scientists than between diplomats, global events and cultural differences can get in the way. In research, trusting another country's investigators may be mired in an internal dialog that asks questions about the collaborators' true intentions. Do they only want access to my data or samples from my country? Will they include our institute in their publications and future proposals? Building trust and good relationships with international partners takes longer and begins more slowly than collaborations with local partners. Memoranda of understanding and agreements, as well as material transfer forms, are important in international collaborations. Treaties and laws have more formalized ways for sharing samples and data between countries, which often leads to bottlenecks or complete roadblocks in scientific collaboration.

When researchers collaborate across borders on global issues that may be messy or political, it can go beyond being challenging. In the scheme of larger political issues, however, scientific collaborations can be a form of scientific diplomacy (in the international relations sense). Using science to bridge two countries' knowledge and technological gaps leads to stronger relationships between them. It is often said that "scientists always get along if they are in the same room and there is a whiteboard nearby." International cultural differences can be considered one of the biggest bricks in the walls that divide us. But vulnerability, trust, and empathy can shatter the bricks that make walls.

Let's envision a scenario in Gainesville, Florida. Jason Blackburn heads off to the airport to pick up an international collaborator who knows little English. The visiting researcher will be there for a few months, or even a year. Blackburn welcomes that person into his family and life. He's starting to fill the trust bucket, drop by drop. Together, his team will tackle a few tough questions to better understand a zoonotic disease. The team will then share the results, most of them in leading scientific journals. Many of the results will indeed be transformational and shift paradigms in the field. And, in some way or another, the results will help make the world a better and safer place.

Exercise

Here are some questions to ask yourself and to think about. You might want to try writing down your thoughts in a journal.

1. Can you think of a time when you had zero empathy for a colleague? Did you change your mind later on, when you understood where they were coming from? Or maybe never?
2. Has there been a time in a collaboration or project when you felt vulnerable? Did you share that feeling of vulnerability, or keep it to yourself?
3. Can you think of a time on a project when you did not perform at your best? What were the circumstances that led to this?
4. What are some biases you may have about differences you have experienced between yourself and colleagues?

CHAPTER 9
DREAM TEAMS

> We shape our buildings, and afterwards our buildings shape us.
>
> SIR WINSTON CHURCHILL

Sir Winston Churchill gave a speech in the House of Commons on October 28, 1943, about replacing the bombed-out House of Commons chamber. "On the night of May 10, 1941," he said, "with one of the last bombs of the last serious raid, our House of Commons was destroyed by the violence of the enemy, and we have now to consider whether we should build it up again, and how, and when. We shape our buildings, and afterwards our buildings shape us. Having dwelt and served for more than 40 years in the late Chamber, and having derived very great pleasure and advantage therefrom, I, naturally, should like to see it restored in all essentials to its old form, convenience, and dignity."

While Churchill was talking about rebuilding after the horrors and destruction of World War II, the quote can apply to many things in life, including teams. We shape our teams; therefore, they shape us. How we build our collaborations, and how much thought we put into finding the right skills, expertise, and attributes, will pay off in the long run. We may ask ourselves, "Is skill and expertise more important than character?" In certain research fields, having the right technical expertise is often required to complete highly novel and extremely specific work. At the same time, team members with a lack of the right character traits will bring down the team and limit the potential of the collaboration. We learn so many things from our colleagues besides technical facts and new types of analyses. We ascertain how to handle stress and adapt to failures. We may absorb how to communicate at all levels and explain the nuances of our work to various audiences. We may discover how to keep positive when facing adversity and know when to stand up and fight the battles worth fighting. We may grasp how to have empathy for a colleague, because we have received a gentle and kind word when facing a hardship in life. We may also learn these things through negative examples, such as when we hear a harsh word uttered or criticism given, or a colleague withholds information from us or speaks poorly of a team member for no good reason. These are ways in which our teams shape us.

Life experiences color not only how we see the world, but also how we approach problem solving. If, when young, you started out working on a diverse team, whether the experience was good or bad, it will modify your future interactions. Understanding how our coworkers "roll" means ascertaining how they prefer to communicate or to manage their time. Each individual on a team will shape how the team rolls. We may find out and accept that a person on the team is a gay man from New Hampshire, but it's more relevant to know if he is a morning or a night person. Or does he get super excited about whiteboards?

Over the years I have heard researchers say they don't become friends with their postdocs or fellow research colleagues. Hearing this made me feel sort of weird and hollow. While a decision to remain aloof from some colleagues may be one I can respect, a US survey from Olivet Nazarene University found four out of five respondents had at least one friend at work. On the flip side, there is truth in the old adage that your manager is never your friend.

ASPECTS OF A DREAM TEAM

When we think of dream teams, we often think of the 1992 US men's Olympic basketball team. It was the first year that the Olympics could include active professional players from various sports. More than 15 years later, the 1992 gold medal team was inducted into the Basketball Hall of Fame as the greatest collection of basketball talent on the planet. It has been said that the real legacy of this team was not that it that dominated every game and won the gold medal, but that the team changed the game of basketball. It showed the rest of the world how to play. In the same way, the hantavirus team showed the world how to respond to an unknown outbreak and ask the sequential questions necessary to understand a disease system. And they added expertise, as needed, along their journey.

How do we build dream teams? This has been a primary question in sports, business, creative endeavors, and science. It does not have to be a Nobel-prize-winning science team, but simply a better team than we would have had without actively trying to build a highly functional group of individuals. In other words, how can we keep moving the project team toward greatness?

As part of the background research for this book, I asked hundreds of researchers of all ages the question, "What are five adjectives or words that describe a successful scientific collaboration?" Over a hundred different answers were given. While the four most common words were "trust," "communication," "productive," and "respectful," a wide array of adjectives were used to characterize great teams. Another of the top ten words was "friendship." When people become friends, they learn about each other and discover how each person rolls with stress, conflicts, and life. Dream teams are where friendships, but not cliques, form on the team, and the success of the team as a whole is everyone's primary goal.

Going back to New Mexico in 1993, we can say that the hantavirus outbreak team was a transformational dream team. They had their arguments and various ups and downs, but they supported each other and were focused on the challenge at hand: finding the source of the unknown disease outbreak in the region. After the outbreak ended and answers were discovered for its cause, the team stayed together to become the RAMBO group and jointly worked on a wide array of scientific issues. People often ask the initial members of the RAMBO team what made them so special. The answer is the diversity of the team, the trust built over time, the bigger mission of working on public health issues, and their commitment to discovering how to be the best team through organizational learning.

DIVERSITY

Dr. Scott Page, a professor of complex systems at the University of Michigan, often says that "diversity is talent." Page's studies have mathematically shown that diverse teams outperform other types of teams. A team can have greater depth and breadth only if their members possess cognitive diversity. We are inherently different in how we think, which is affected by our experiences and history, as well as by the identities we may have for ourselves, and those we acquire through how we have been treated by others. Our identities may be race, religion, gender, the community we grew up in, or even how we identify as a scientist. There are millions of possibilities. These self-identities and life experiences have led to differences in how we think. This includes our diverse mindsets, problem representations and problem-solving skills, knowledge bases, heuristics, technical and communication skills, and stress responses to situations. In his studies, Page showed that these differences are what enable teams to find more novel and creative solutions, make fewer technical or conjectural errors, and develop more accurate predictions. His book emphasized that he was making a practical case, different from the broad cultural narrative stating, "Diversity is important." Page also stressed that random diversity is not the answer for great teams. He argued that the character of the problem shapes the type of diversity required. We can see this in the hantavirus team, when they continued to add technical expertise to help solve the evolving questions that shifted as they learned about the cause of the outbreak.

In his work with understanding the role of diversity in a team's success, Page categorized individuals' cognitive repertoires into five facets. The first is cognitive diversity, which has to do with information and what data and facts we may know. The second is knowledge, which is expertise in a particular domain. The third is heuristics, which consists of the techniques we know and those we discover and apply. The fourth is representations, which are the perspectives we bring to a situation, based on our background and experiences. The fifth is the mental models, or beliefs, we may hold about how the world works. Such models could either be simplified or systematic descriptions of an understanding. The group's repertoire will be the union of the individual members' cognitive repertoires. Page also stressed that team diversity is not like portfolio diversification, which spreads out risks and gives an average outcome. It is about having the right diversity for the team at a particular time and situation.

GOALS

There are many ideas on how to build and develop a dream team. In science, it is common to think it is built by bringing together top people who possess sufficient skills in the complementary fields that are needed to answer the question at hand. Yet it is also important to set individual and team goals, as well as to build on individual strengths. Team goals are often clarified and outlined in the proposal, but individual goals are often not as clear once a project starts. For example, a team goal could be merging two complex models or designing a large laboratory experiment. The individual goals on a project, however, may vary greatly. These diverse goals may be to acquire computer hard-drive space, or make sure that the laboratory is compliant with biosafety regulations, or see that the sequencer is serviced and ready. There is a common saying for leaders: "Set the banks of the river and let your people flow within them." This means that an ideal leader will establish individual goals and expectations, and then step back and help at the project level, with less interference at lower levels. Equally important is setting the project's expectations and scope to match the abilities of everyone working on all parts of it.

THE WORK CULTURE

Most of the articles written about building a dream team stress the power of a positive work culture. All levels of its members can create opportunities for the entire project team to take part in together, thus strengthening workplace relationships. This was critical for the hantavirus team as they developed and grew the RAMBO group. Insights into their current research projects were often shared and discussed, not only in formal project meetings, but also over dinner or other recreational activities. As discussed in chapter 1, the RAMBO team studied and applied the organizational learning theory of Peter Senge, using his book *The Fifth Discipline*, initially published in 1990. They strove to be a learning organization. The team sought to understand how everything fit together and how every aspect of the team was interconnected.

PROTOCOLS

Another top determinant of dream teams is their protocols. To be effective and successful, teams need protocols—that is, a detailed set of conventions or plans that establish ground rules for standards and behavior. While creating codes of ethics has become more common in scientific societies and institutions, developing protocols for projects is a new trend. Protocols, or expectations, must be discussed and accepted in advance, to avoid miscommunications and misunderstandings. This includes defining everyone's roles and responsibilities, and setting rules for team interactions that foster a positive work environment. Previously, project leaders might have let a joke or a microaggression go by without any comment or negative consequence. It is now more common to discuss the expectations for how we treat each other when projects start or as issues arise. I was recently on a team where it was brought up that a team member talked over what women and the group's younger members were saying. This was then discussed—as a team—in a conversation that both described the expectation to not talk over others and asked if anyone had thoughts or ideas on this topic. The problem was addressed immediately and directly, without fanfare, but it was addressed, rather than ignored.

SELF-AWARENESS

Dream teams start with the leader. One of the best ways to be an exceptional leader is to have self-awareness. This means paying attention to and being aware of the different aspects of the self, including your negative and positive traits and behaviors. In *The Science of Dream Teams*, Mike Zani provided a great framework, which he called "front of shirt, back of shirt." Visualize the front of a T-shirt, listing all our accomplishments and positive traits. The back of the shirt also lists things about us, but these are what people might think of as our more negative traits and behaviors. Maybe you tend to interrupt conversations, or seem to have obvious biases against people. Problems arise because we can't see the back of our shirt. This creates a gap between how you perceive yourself and how others see you. I would add to this framework by noting that sometimes how we perceive ourselves may be more negative than the way in which others see us. This might be particularly true for women, minorities, or anyone in science who subscribes to an imposter syndrome—that is, who doubts their skills, talents, or accomplishments. Finding out what is on the back of our shirts can help us improve and fix our bad habits. It can also help boost our confidence in ourselves when the back of our shirt has a shorter list than what we envisioned about ourselves.

There are many ways to find out what is on the back of our T-shirt, besides the Tina Turner or Van Halen tour dates for 1984. The most direct and comprehensive way, one which is common in business, is known as a "360° review." In it, an assessor reaches out to all your colleagues and asks specific questions about your strengths and weaknesses. Their responses are anonymous. I once had a male colleague who had a 360° review completed as part of a leadership mentoring session. Based on his interactions with the team afterward, all of us must have answered that we could not trust this person, because he literally had zero trust in others. A week or so afterward, he casually brought up in a meeting that the reason he did not trust people was because he was trained as a scientist, and that, in science, you are taught to not trust things and get the data yourself. Yeahhhhh, but it wasn't really that. Self-awareness for an exceptional leader means accepting the negative things about oneself. This can lead to growth and to working on correcting poor behavior traits.

Another way to become more self-aware is to undergo one of the dozens of different psychology and behavior assessments that are available. Some may argue that these assessments have no value, due to generalities and a lack of concrete ways to scientifically gauge how true the assessments are to whom a person really is, but I would differ, believing that they do have benefits. These assessments help train us to focus on asking questions about ourselves and how we work with others. The answers and the categories applied to us, based on answers from an assessment, can also help inform us of our traits in general, both the ones we want to highlight and the ones that we, perhaps, want to work on improving. Another way to find out what is on the back of our T-shirt is to ask for direct feedback from our colleagues. This can be from our students, research peers, and leaders in our organizations. Feedback in the workplace is a valuable, yet limited, commodity.

Leaders of dream teams have exceptional self-awareness. We can all think of star principal investigators that lead top laboratory groups, receive millions of dollars in NIH funding, and have storied careers—but are also known to be jerks. Nonetheless, in science, and our society in general, the days of domineering and harassing leaders are rapidly changing, and, as many would argue, ought to be over.

HOW WE TREAT OTHERS

Society is at a crossroads right now—reinventing itself toward diversity and inclusion, while at the same time losing empathy toward others. This might explain much of the stress people currently feel in their workplaces. Scientific societies are now judging researchers for awards not only on the basis of academic accomplishments, but also on leadership and their support of the communities they work with. News outlets are now filled weekly with articles about a top researcher's fall from grace through sexual harassment or misconduct. Many of them tell of toxic work environments in the laboratory. While research demands excellence and hard work, these stories can be examples of how noncooperative collaboration did not improve the science. The switch that is being made in science—and the premise of this book—is that how we treat each other and work together leads to better and transformative science, and does so faster.

How we treat each other includes how we treat students and early career professionals. Similar to parenting, many believe that being hard on students is in their best interests and teaches them to have a thick skin toward the hardships and competitiveness of research. But having high expectations of students and challenging them is not the same as shaming them in order to make them tough, as a way to excel. People push themselves out of fear, as well as from not wanting to let down those they respect. A domineering leader of a laboratory who threatens dire consequences for not meeting expectations may get great results from people, but this practice is not sustainable in the long term. A respected leader who leads by fostering the inclusion of everyone on a team will also get great results, and this technique will be more sustainable over time, be more enjoyable during the journey, and allow everyone in the group to grow into the best possible version of themselves.

There are also many stories of leaders in science who may show respect for their project's peer collaborators but treat their students and lower-level laboratory members with disdain or shame. I was once in a meeting with a very well-known, established scientist and his two much younger postdocs. In this context, I was acting as a science program manager with another of my colleagues, and we got together to discuss the possibility of funding this laboratory group. The famous scientist proceeded to tell us that he would write the proposal himself, composing it on a flight he was going to take, because "his postdocs could not write themselves out of paper bags, and [were] horrible writers." The disappointment and shame on the faces of the two postdocs was palpable, and all I could feel was embarrassment for these younger yet very accomplished researchers. After the meeting, one of them met with us and apologized for the lead scientist's remarks. That postdoc expressed great interest in the project and mentioned having wanted to try writing the proposal first. Needless to say, both postdocs soon left this scientist's laboratory and are now engaged in trying to break the chain of treating younger people in research poorly.

Learning how to better communicate our frustrations with colleagues at all levels can be challenging, to say the least. If a colleague fails to deliver a deck of important presentation slides, or a laboratory experiment is done in a haphazard way, anger can be an appropriate response. How we navigate being able to not repress that anger, and to let others know we are angry or disappointed, while doing so in a constructive way, is the emotional intelligence we all hope to have on any given day. Leaders of dream teams that continue to stay together over time are the types that might read a book like this one, or other self-improvement treatises for the workplace, such as Stephen Covey's *The Speed of Trust*. Leaders that grow and evolve also seek to find out what is on the back of their T-shirts. A leader's job is to make their colleagues and collaborators stronger and better.

MOTIVES FOR COLLABORATION

Research teams are developed differently from teams in a traditional business environment. The former are designed more by the team members themselves and the principal investigators. In three different studies and in the book *The Strength in Numbers*, Bozeman and his colleagues asked researchers what their primary motivations for collaborating were. The top answer was "working with researchers whose skills and knowledge complement mine." The second most important factor was "increasing my own research productivity." Two other leading factors that motivated people to collaborate were "working with persons I can depend on to complete work on time" and "having fun working with researchers I like on a personal basis." For each of these responses, I can think of colleagues who complemented my skills, and colleagues I can always depend on. I can also vividly remember moments of laughter and connection. One might be the time when a colleague bent over to pick up a chicken and their scrubs spilt right up the backside. That person was completely nonplussed but remained committed to finishing the sampling. I and others can probably also recall some of the other million ways when the science did not go quite as planned, and all we could do was laugh about it and move on.

THE CONSULTATIVE MANAGEMENT APPROACH

While there is no one best way to solve all research collaboration problems, there can be management approaches to help move a team along the x-axis of research collaboration outcomes toward a dream team (figure 2). Also, there are various research management methods that describe different ways in which good leaders manage their teams. The consultative collaboration management approach was developed and preferred by Bozeman and Jan Youtie. The crucial point for this technique is that "all parties of the collaboration are consulted at key points in the life of the collaboration in order to identify respective preferences and values and decide upon specific actions."

These authors considered the consultative collaboration style to be the gold standard for managing research teams. Key points in a project might include writing its objectives, setting goals, assigning tasks, outlining possible presentations and publications, and developing programs to sustain the project. These are also the characteristics of a consultative collaboration. First, ensure that enough communication structures are used at key points. These structures should include whole-team communications, not just those to subgroups. Second, not surprisingly, create a high level of trust among collaborators. This leads to a commitment to open and transparent disagreements when such situations occur, and it also helps prevent researchers from associating the assessment of ideas with the person presenting them. This is a particularly hard skill to develop if it is not consciously practiced. Third, baggage can accumulate in teams over time, as it does in any relationship. Both conscious and unconscious biases can be part of how we see other people. Relatedly, a consultative collaborative team recognizes diverse values and ideas and is sensitive to the status and power dynamics within a team.

AIRING DIFFERENCES

Marriage counselors will often say that the biggest factor in divorces is not how much partners argue, but when they stop talking. When we perceive differences between ourselves and other people, we often become quiet and less apt to share our opinion on things. There may be power dynamics at play within a team, and we may not speak up. Researchers may feel unease because they are minorities, or in the LGBTQ+ community. A tendency to stop talking can be especially true for scientists who are introverts. I could find no concrete data or study that validates the hypothesis that *most* scientists are bookish, introverted, or shy. But collaboration may be more challenging for those among us who *are* introverts. I once was in a leadership training session where all of us were fellow scientists. We took an introvert versus extrovert assessment. At least for our group of about 20, the results fell as a bell curve, with most people in the group split 50/50 between introversion and extroversion. Only a few were on the edge of being at one extreme or the other. This helps to explain why, at a science conference, I am fully engaged with conversation for about an hour, but then I literally have to run away to hide in a bathroom or empty hallway, figuratively thinking that my hair will catch fire. Those emotions later subside, and I start to feel lonely, so I have to go find some colleagues to engage with again, thus continuing my conference cycle of craziness.

PSYCHOLOGICAL SAFETY

Teams that actively encourage dissent and the airing of different viewpoints in a supportive way are more innovative and do transformative work. This means that everyone on the team—including minorities, women, members of the LGBTQ+ community, and introverts—feels comfortable speaking up and sharing their opinions, even if they clash with some members of the group. This form of dissention is not unusual in teams. It is commonly referred to as "cognitive friction," which drives innovation and transformation. The great leadership guru John Maxwell said that for every interaction with others, including when we collaborate, we unconsciously ask ourselves three questions. Do you care about me? How will this benefit me? Can I trust you? What these questions establish in our mind is called "in-group psychology." This is where our brain divides people into two groups: ones that are more like us, whom we can trust; and the other group, whom we should not trust.

Crucial conversations are those important discussions that lead to consequences in a relationship or topic, or, as the subtitle of a 2021 book on crucial conversations states, when the stakes are high. For a crucial conversation to be successful, participants should feel like they are psychologically safe. Trust is the belief that another's motivations are kind, and that this person has our best interests at heart. Humans cooperate not only for reasons of self-interest, but also because they are genuinely concerned about the well-being of others. In order for social systems to be sustainable, we must balance trust with distrust, but mostly trust others. Yet when the stakes are high, our emotional gauges are also set on high. Thus feeling safe in any interaction, including our most intimate relationships, can change instantly. This alteration can happen when a negative feeling may be triggered, or a topic is brought up that you totally do not even want to think about, let alone discuss.

How do we create a safe place to have a discussion? One reason we may feel defensive when someone brings something up is not that we do not like the message, but that we believe their motivation or intention is pernicious. Because of this, the first condition of safety is ensuring a common purpose. Pointing out mutual intentions and goals helps show that you are on the same side, whether you are moving toward a solution to a problem or trying to develop better communication. Something else we can do to create a foundation of safety is to remind ourselves to have respect for others. This may be more common in science, but there are still situations where one researcher may actually not respect another. Even if they never admit it, that disrespect comes through. It might be that they believe respect has to be earned in science. While our skills and expertise increase with experience, often the ideas of younger researchers still have value and help us to see a situation differently. Respect for others is an important part of inclusion and equity in the workplace. Establishing respect for others helps ensure a greater level of safety and trust.

Issues and challenges that our society is dealing with at any moment find their way into our laboratories and projects. This may occur in the case of a transgender researcher, a graduate student who is the first in their family to attend college, a Black student who feels unseen, a colleague who may let their political views be heard more loudly, a researcher who engages in inappropriate sexual innuendos or behavior, or a colleague who does not want to get a COVID-19 vaccine. These situations and beliefs can impact communication and connection more deeply than the more usual differences we come up against in a project. In my decades of collaboration, I have experienced all of the above, and I expect most researchers with even a year or two of collaboration under their belts have already encountered many of these, as well.

PUTTING IT ALL TOGETHER

Dream teams seldom have individuals in the group who are not pulling their own weight or are not engaged. There is a collective vision that is clear to everyone, and a finish line is in sight. For the 1993 hantavirus team in New Mexico, the initial finish line was identifying what was killing people, finding out how they got the disease, and then being able to stop the outbreak and its further transmission. The next finish line was to discover what the bigger causes of the outbreak were, understand why it happened, and predict when it could happen again. Relatively speaking, the team crossed both these finish lines quickly. Although the team grew to include members with new skills and expertise, everyone still stayed engaged, for two reasons. First, everyone was aware of the significance of the outbreak, and that current and future lives were at stake. Second, the team was a learning organization, committed to being the best team they could be, and established trust among its members daily.

The hantavirus team had both vision and purpose. Vision is a specific destination, a clear picture of the desired outcome. In this case, the desired outcome was identifying the cause of the outbreak and discovering how it happened. Purpose indicates a general direction, and thus is more abstract. In this case, the team's purpose was twofold. First and foremost, it was to stop people from dying. Second, it was to then create a phenomenal group of researchers in the desert of the Southwest, who came together to support each other and better understand infectious diseases in the region. Purpose is what allows a team to set new visions as time goes on, and it can lead to excellence. Vision is the specific goal at the end. You need both to be a dream team.

Once the hantavirus team moved on to their next purpose—creating an outstanding network and team in RAMBO—they studied Senge's fifth discipline: getting better at supporting each other as a team. One of the core constructs in Senge's book is the idea of creative tension, where disagreements ultimately give rise to better ideas or outcomes. Creative tensions are also the gap between your vision and reality. Think of these tensions as a rubber band between the two disparate aspects, where you can do one of two things. You can pull your vision toward reality, or vice versa. In terms of the disease outbreak, the hantavirus team pulled reality toward their vision by adding new skills and fields to their group of researchers, to better understand what that reality was. For the duration of RAMBO, the team studied how to master creative tension, in order to be more disciplined and bring out their capacity for perseverance and patience. Playing the long game by researching the environment's role in outbreaks and climate change meant the possibility of many proposals being rejected, multiple ideas being tested, and decades of hard work. The team wanted to enjoy the ride together, as well as support each other as individuals. To them, time was not an enemy, but an ally.

In a way, the science of team science is the next fifth discipline for science. Technologies, computers, analyses, and engineering have attained such a level of excellence that improving science is now about improving how we work together. People often say that you write a book for your younger self, containing information you wished you had known earlier. I can remember that when I was doing research for my master's thesis, I had a terrible fight with my assistant. We both felt we were right about something, although I now don't remember what it was. What I do remember clearly is that we were both walking through a grassland in the wilds of North Dakota, carrying shotguns. We stopped midfield and became acutely aware of the absurdity of our argument, and then bent over laughing. Later, I realized that I did not like conflict with my colleagues while doing the research work I loved. After the fight in the field that day, I did mellow out. I stopped taking myself so seriously, and I never again had such a contentious argument in a work environment. Or while carrying a shotgun, for that matter. At this point, I also started reading books about business teams, communication, and leadership, in order to learn how to improve my interactions with others and become a better scientist. Maybe this was partly because I felt like I was an imposter, trying to make up for a lack of skills or intellect. What happened on this journey was that I did become a better scientist. And what I primarily discovered was that what I loved most about being in the research field was collaborating with my colleagues and spending time with them. I listened to my colleagues, I learned from hearing different viewpoints, and I met amazing scientists from around the world. I have enjoyed so many dinners filled with conversations and laughter, and I have shared many cups of coffee and beers with them. I have stood at a whiteboard with my colleagues as we hashed out our sampling plan, paring our proposed 25,000 samples down to 100 after looking at our budget. I have been a part of extraordinary teams, and some of them might even qualify as dream teams.

Exercise

Here are some questions to ask yourself and to think about. You might want to try writing down your thoughts in a journal.

1. When you think about a project team you have been a part of, what is your opinion of the excellence of that team? What are the strengths and weaknesses of the team as a whole?
2. What are five adjectives that best describe a successful collaboration?
3. Can you think of a difficult or painful conversation in the past, occurring in a workplace setting, where you were a participant? What made it hard? Who was the instigator?

CHAPTER 10

SCIENCE NETWORKS

> Our job in our lives as we walk around our world is to create a space around us that invites people to have a better relationship with us.
>
> KEITH FERRAZZI, *author and entrepreneur*

Our social networks help sustain our need for connection with others. Most of us have heard about and experienced the power of social networks and how they can shape ideas and beliefs, as well as connect people globally. Everyone you work with and know in science is part of your bigger science network, ranging from your administrative assistant, who helps navigate the complicated governmental system to submit proposals; to the journal editor; to the janitor who cleans the laboratory. Each one plays an important part in keeping research moving forward one day at a time. Networks can be informal or formal, small or global, focused or broad in scope, and engaging or not.

The definition of a network—outside of the science of team science, where it is important to have standardized definitions—can be stated in many ways. Networks can simply be another word for a collaboration. Or they may be how different projects can connect with each other. For example, a scientific society is a network. Or it may be a collection of organizations, such as the Societies Consortium on Sexual Harassment in STEMM (science, technology, engineering, math, and medicine), which was developed to set standards of excellence in STEMM fields for high-quality research and professional and ethical conduct. Science societies have joined this consortium to gain expertise concerning ethical standards, such as access to tools to deal with the challenges of bad behavior in science, like sexual harassment. This knowledge is then passed down to and benefits individual members through codes of conduct and ethics, training sessions, and heightened expectations for how members behave, both at meetings and in their professional life in general.

Formal networks take time, energy, and resources to maintain over the long run. Some networks may start out formally and can later become more informal, requiring fewer resources once they are established. The COVID-19 pandemic has taught us many things, and one of them is the importance of our networks, which keep us going and give us support. The pandemic forced most research collaborations to slow down, pause, or stop altogether, depending on the situation. Collaborations that used to have in-person meetings pivoted to virtual ones. Laboratory work and field experiments were put on hold. Science conferences moved to online venues, and poster sessions were presented in Zoom, minus the beer and snacks. We have met new people through our networks and now have relationships that are completely virtual, allowing us to comment on each other's pets when they enter the room or ask where the background photo was taken.

The infectious disease group from the nation of Georgia, mentioned in the previous chapter, is involved with several formal networks, as well as being part of a large, informal, global network of infectious disease and public health experts. If you work with the Georgians and have an in-person meeting in their country, when someone asks what the trip was like, you probably would be inclined to go into a deep discussion about the food you ate, ranging from the cheesy bread called *khachapuri* to *khinkali*, which are meat dumplings. And then you could move to an analysis of why Georgian wine is so different, how that country is the cradle of wine, and how archeologists have traced the creation of the world's first wine, in the southern Caucasus, back to 6000 BC. After about 20 minutes of listening to your reminiscences, the questioner probably will rephrase the question to say they meant, "How was the research meeting?" You then might enthusiastically state that the meetings were great. You got many kinks worked out in the sampling design and plowed through the analysis of the first year's data. You might then add that at the first data analysis meeting, the team ate *churchkhela*, which look like candles but are actually shelled walnuts that are strung together on thread and then repeatedly dipped into a thick grape juice syrup known as *tatara*. The walnuts are air-dried and then dipped in *tatara* over and over again, until they are coated with a waxy, fruity buildup. The string is then pulled out, and the candle-shaped lengths of coated nuts are cut into bite-sized tasty snacks, which were originally created to give soldiers energy as they headed into battle. They're also perfect for a long session of writing standard operating procedures or performing data analyses.

The Georgian team is part of the Western Asia Bat Research Network, or WAB-Net (pronounced "wabi-net"), designed to promote collaborative bat research across western Asia. It is part of a group of threat reduction networks, first established by the CTR program discussed earlier. These networks are a collective of individuals, groups, and organizations with a common goal—reducing biological threats. Their mission is tied to biological issues: security and safety, surveillance, and cooperative research. WAB-Net is made up of seven initial countries, including Georgia, that benefit from the shared experience of collectively testing hypotheses and learning new techniques that strengthen their capabilities for biosurveillance.

These types of research networks are designed to foster multidisciplinary approaches to large challenges and questions that science is trying to investigate. This is particularly true for the One Health approach to understanding diseases in the context of humans, animals, and the environment. It was also the approach used by the first hantavirus team, before the term was even coined in 2009. The hantavirus team had to take multiple factors into account: the humans getting infected and their families' living conditions, the small rodents that carried the disease, and the environment that helped lead to the outbreak in 1993, plus ones that might occur in the future. Regional research networks can enhance communication between countries, or states within a region, and they can bring together potentially opposing fields, such as conservation and responses to infectious disease outbreaks. Designing a network that includes behavioral and social scientists and economists is a hallmark of a One Health approach—or it should be, as most problems are sociological in nature.

During the mid-1990s the National Science Foundation began to talk about the need to foster more activities that connected different fields, as science was moving toward more multidisciplinary efforts. Dr. Sam Scheiner is a biologist and program manager in NSF's Division of Environmental Biology. In 2001, Scheiner and his colleagues wanted to create networks of scientists working together, but who were not based within a shared scientific facility, such as a center of excellence. Research Coordinated Networks, or RCNs, were born from the seeds of these ideas. The stated goal of the RCN program is to "advance a field or create new directions in research or education by supporting groups of investigators to communicate and coordinate their research, training and educational activities across disciplinary, organizational, geographic and international boundaries."

While the RCN program has been immensely successful in fostering new collaborations and innovative ideas, it is somewhat altruistic for researchers. RCN funds are spread across the network and directed more toward annual meetings that are intended to bring everyone together. RCN proposals are similar to normal research proposals, although there are many more researchers to communicate with, in order to capture their ideas. Every RCN project is unique and may be scoped around any topic, such as an emerging field, like ecoimmunology, or a taxon group, such as Manakin birds.

When I asked Scheiner for his best advice in creating successful RCN proposals and projects, he responded by giving equal weight to the management of the project, as well as the science. He suggested having clear goals and objectives, described through a management structure, and a premise that is inviting not only for new people, but also for new ideas. He also mentioned focusing on what should happen between the in-person meetings, in order to keep up the momentum. I next asked him what some of the challenges were in leading or being a part of an RNC. He responded that leaders in a field may not be invited to join an RCN, because those with big egos often get passed over. Science networks are about collaborating—moving ideas and innovations forward through teamwork. A reputation for being egotistical or difficult to work with can supersede all other factors, as researchers who love working with other researchers may not want to intentionally add conflict. On the other hand, it is important to include researchers who have different or novel ideas. When asked, most researchers say they know the difference between exhibiting novel or forward thinking and being a jerk.

One of the goals for any form of coordinated science network is to prepare the next generation of scientists and research professionals by attracting more young people to and retaining them in STEMM careers. This includes widening the participation of all types of people in science—that is, focusing not only on individual diversity, but also on geographic regions, types of institutions, and new disciplines. In broadly diverse teams, communication becomes even more important, and having a good communication plan is essential for keeping momentum going within a network, as well as building trust over time. Having the right balance between how often those in a network communicate with each other and the types of communication that take place is vital for sustainability and success.

Dr. Erik Olsson is a theoretical philosopher and computer scientist at Lund University in Sweden. For him, group inquiry is an essential part not only of society in general, but also of scientific investigations. Due to instant electronic communications, group activity and interactions have increased dramatically over the past three decades. Olsson thus asked himself two questions. Is communication beneficial in science investigations? And does it actually result in better conclusions? Since a majority of the present book is focused on increasing communication, these are important questions to ponder. Group inquiry is designed to focus on one or two questions, to be solved by collective means, rather than look at multiple subquestions, which can lead down a rabbit hole.

As Olsson set about answering his question of how useful communication is for solving a research problem, he uncovered studies in social psychology and economics which found that some forms of communication links, such as network density, may be detrimental to group performance. Olsson created a software program, Laputa, to test his theories about research networks, broadly defined. Laputa is a research tool and computer environment for simulating the attainment of knowledge in social networks, such as online networks, scientific departments, editorial boards, and expert groups. The basic premise of Laputa, outlined in several publications by Olsson, is centered around two primary concepts: inquirers and links. As Olsson stated, "A society consists of a set of inquirers, and a set of communicational links between these. Each inquirer determines how she is to update her degree of belief in the face of evidence, and how she is to find this evidence. The links determine which inquirers can receive information from which, and how this information is handled. Thus, a link to an inquirer can also be seen as an aspect of her investigative behavior."

What Olsson and his fellow Lund University collaborator, Dr. Staffan Angere, found was that unrestricted communication in a network was a bad approach to converging on the truth. They used two primary metrics to measure the impact of a simulation on how well a network responded: the notion of a veristic value, or V-value, and the average belief of the majority. What is it that makes network members less likely to converge on the truth? The short answer is that casually asserted information inserted into the network can pull the entire network away from truth. The take-home message for researchers is to show discretion concerning what information is shared with the network and to use proper restraint in your communication behaviors. Olsson and Angere suggested that this also applies to scientific publishing, in that the greater the number of published research articles that contain doubts or lesser-quality designs, the more the research field in general will veer away from truth. Just as we must take care in validating what is posted to social media, the same is true in our science networks.

In New Mexico, Gary Simpson helped me carry a U-Haul moving box to my car, and it took both of us to lift it into the trunk. The 41-pound cardboard box was filled to the brim with newspaper articles and notes on the hantavirus pandemic, dating from 1993. The sources for the clippings ranged from the *Navajo Times* to the *New York Times*. Now almost 30 years old, they contained headlines such as "Of mice and men: Residents on their own" and "Desert Mystery: Death toll at 13." There were reports of the twelfth victim going downhill in just two hours. The news articles covered the gamut, advancing from documenting two cases of Four Corners flu, to describing a mystery plague, to reporting over 42 confirmed hantavirus cases over the summer. They traced the story from unknown causes to hantavirus to rodents to deer mice. By June 22, 1993, news sources were carrying the most accurate information available on what was known as a "biological serial killer." One of the best headlines was, "Disease trackers use classical logic, computer cross-checks, mountains of data and true grit to hunt down killer illnesses."

During the summer of '93, debates on how to deal with the outbreaks were on the front pages of newspapers. The State of New Mexico announced that it would not try and kill off the deer mice, as they reproduce too fast. The impacts of coyotes and coyote killing were discussed, with the breakfast contents of 19 mice in the stomach of a typical coyote being offered as evidence of the positive role coyotes play in the ecology of the Southwest. There were articles documenting ancillary effects of the disease in the region, from drastic declines in the number of tourists to sales of Native American rugs drying up. On the other hand, there were also repeated articles on the miracles of bread machines and "the men that loved them." To combat deer mice on their property, people started bringing home cats, and then the CDC expressed increasing concern about the cats transmitting plague. On November 20, 1993, the *Albuquerque Tribune* reported on a hantavirus breakthrough: the CDC had successfully isolated and grown the virus. There were editorials and full articles arguing that the virus had to be a result of germ warfare. By October 1993, a Nevada firm was marketing a "hantavirus home kit" for just $19.95, which was an iodine-and-detergent solution to knock out the virus. Members of the hantavirus research team went on record to say that this kit was not worth your money. By October 29, 1993, the CDC renamed the Four Corners hantavirus, now calling it "hantavirus pulmonary syndrome." At the same time, they confirmed that the first hantavirus death was in July 1991.

The hantavirus team's careful comments were tracked daily in the papers. Cautious researchers stated that rodent droppings were a suspect in virus transmission. New breakthroughs happened each day. The team was transparent with their findings and open about what was still not known. There were discussions about the process of science, comparing it to the headlights of a car on a dark road, only showing brief glimpses of the way ahead. On June 19, when researchers reported with confidence that they had finally identified the rare virus, there was a rally held at Window Rock, Arizona, to demand a halt to focusing national attention on the Navajo people, as it created a fear of Native Americans and further deepened the divide with other Americans, who literally ran away from them on sight. During the outbreak, the New Mexico Department of Public Health created a hotline, where staffers fielded questions about everything from pet care to devil worship. In a news article, it was facetiously referred to as "1-800-QUIRKY."

By mid-June, there were highly detailed graphics in the papers, explaining how the polymerase chain reaction test worked. At the same time, editorials were being published, sporting titles such as "Dr. Gary Simpson: Successful Medical Sleuth or an Incompetent Pompous Ass?" This one accused Simpson of developing the "Gallo-Fauci syndrome," referring to the work of Drs. Robert Gallo and Anthony Fauci in identifying the AIDS epidemic. If you are beginning to feel a sense of déjà vu, then you would be correct. It is against this now-expected backdrop of chaos that scientists must currently function in any public health or other crisis. It can no longer be a surprise to us that the public will misinterpret the facts, as well as what researchers are saying at the moment, about whatever is known and unknown.

What is amazing about the 1993 hantavirus team is the sustainability of their efforts, which continued for over 25 years. The productivity of the team that went on to describe a One Health system of disease ecology was incredible. Not only did their science undoubtedly save many lives over the years, but the team's work also expanded into many other advances in their respective fields. In addition, their lives were enriched beyond measure by their friendships and their mentorship of the next generation of young researchers. The remaining members of the hantavirus team and RAMBO network across the American Southwest say that they worked hard at creating sustainability, studying how to be a learning organization, and continuing to grow. Like a marriage, collaboration takes work. That effort can sometimes be messy, not perfect, emotionally painful, and annoying. But it can also be joyful, funny, insightful, and emotionally filling, leading to individual growth for everyone involved. A team makes that growth easier, particularly when it is echoed by the community as a whole.

One of the growing challenges that science and researchers now face is mis- and disinformation campaigns against the most well-established data. These campaigns can be created about and spread against science fields in general, and even individual scientists. Researchers are left with not knowing how to respond to misinformation. And disinformation can now include erroneous data that is inserted into shared databases in science, such as GenBank.

Exercise

Here are some questions to ask yourself and to think about. You might want to try writing down your thoughts in a journal.

1. If you could design any network of researchers, what would the topic be focused around? A subfield or methodology? An emerging challenge or question?
2. If you could go to dinner with five scientists, whom would you invite? Would it be a diverse group of mentors to whom you would ask questions? Or would they all be in the same field?

3. Who is one researcher you would like to meet and get to know? Is there anyone in your network who is acquainted with that person and could introduce you?
4. Are you the type of person who likes small research teams? Or do you prefer larger teams, with many different people you could get to know?

CHAPTER 11

WHAT THE HELL JUST HAPPENED?

> Nothing in life is to be feared, it is only to be understood. Now is the time to understand more, so that we may fear less.
>
> MARIE CURIE

The door slammed behind a researcher who had left the room to get some air and take a break from the project-planning meeting. Not listening when colleagues tried to ask what was wrong, the disgruntled individual made vague gestures in the air and said, "I'll be back later." The two colleagues who were left, one of them standing at a whiteboard, blinked at each other in disbelief. Each was thinking, "What the hell just happened?" Before then, everyone in the group had great rapport and called each other friend. But on that day, the team circled round and round on the same topic. One member grew more frustrated as the minutes ticked by, while the two others thought everyone was on the same page. The communication was so bad that these two did not even realize there was a disagreement, which only became evident when the researcher who stalked out stated, "I'm tired of being told I'm wrong." It seemed incongruous to the other two team members. How could they have told someone they were wrong when it did not fit with their own experience, which was that everyone seemed to be in agreement? Why was one researcher so frustrated with the conversation? Why were the team members not in accord with each other? In other words, "What the hell just happened?"

Many scientists may relate to the above situation, with a conversation seemingly moving the project forward and the next statements making it go around in circles. This circling is so slow and innocuous that when the discussion comes back to a particular topic, there is a déjà vu feeling about it. Then, with each additional passage, frustration grows and hopelessness sets in. Suddenly the stress boils over in a team member, and the meeting must pivot to stop the insanity. Some members might believe that if they keep repeating the same thing, the other person will finally get it. Others will not feel as though they are heard and wish they were not in the meeting at all. And some members may be thinking about weekend plans and not be engaged at all.

The above example of a frustrated researcher touches on many of the topics discussed in this book. Did this person not trust their colleagues? Was there a miscommunication, or an expectation not discussed and agreed upon? Was there a stressful situation for the disgruntled researcher outside of work? Was it way past lunchtime? Were there team members who were not completing tasks or breaking promises? Finding the answers to these questions may first require creating an atmosphere of psychological safety, following up after a disagreement—or a blowout—by asking questions, and then validating that all team members thought they were heard. Validation does not mean that you have to agree with each other, but it does show that you have understood and acknowledged all of the ideas and opinions expressed.

Later, when the first researcher returned, the remaining two investigators—although tempted—did not start by asking, "What the hell just happened?" Instead, one of them stated, "I can see that you are frustrated and that must be difficult. Are you feeling better, and would you be willing to share some of your frustrations? We would love to better understand the difficulty so, together, we can work on the issues. We did not mean to upset you with anything we said." The upset researcher smiled, sighed in relief, and then proffered an explanation. As a group, everyone did not understand the expectations of each of the team members, with them all knowing how they fit into the project. The person who left was not even sure what created such a bad feeling at the time. That individual just needed to get away and collect their thoughts, as the stress was growing and becoming overwhelming. The team then pivoted to outlining the project's tasks and required capabilities and discussing how each member fit into the different parts of the work. Afterward, it was clear that a layer of trust had been added, and they had each other's back.

My journey in learning about teamwork and collaboration in science stemmed from my wanting to make up for my technical or other inadequacies in being a good scientist, and thus sustain my career in research. Seeing colleagues struggle with other collaborators, and having my own experiences like the one above, I wanted to apply what I had been learning from the literature on teamwork and leadership. I longed to learn how to work better with colleagues and be a role model for the younger generation. I collected stories and met with scientists to discuss collaboration and teamwork, always searching for a next bit of advice that could help me be a better researcher. While I learned many things on my journey, such as how to properly validate someone in a conversation, I discovered that what I like most about a career in science is working with others. I also found that the field of science is changing in how we treat each other. Leonardo Almeida-Souza and Lilian O'Brien eloquently argued for a kinder approach to science in a 2022 article, ending with the words, "Our wish is that kindness becomes a habit." My hope is that it may become a very enjoyable habit that gets easier as we practice it more.

When scientists disagree, it does not have to be acrimonious. When a team member is feeling supported, able to bring up any issues and be heard, it can lead to better problem solving in meeting the challenges all projects face. These difficulties can be numerous and could be bigger than anyone ever imagined. Strong collaborations can transcend the events and challenges a team may face. An example of the "new normal" in terms of extreme situations is one of my current projects. It consists of an international team with members coming from four countries, including Ukraine. The team has met biweekly via Zoom. In February 2022, as news reports were surfacing about the possibility of Russia invading Ukraine, our Ukrainian colleagues were calm and seemed not too worried about what was to come—at least until the first night of the bombing that hit multiple cities across Ukraine. One colleague missed the next Zoom meeting, and the team was beyond worried. But soon after that, our collaborator responded with emails to us all, thanking us for our concern and for caring about him in these difficult times. A few weeks later, he was able to join the project meetings again on Zoom (to a cheering team uttering squeals of joy). Seeing all our of colleagues on video has made such a difference.

Throughout 2022, the team learned to operate in the new normal of a continued pandemic, supply chain issues for laboratory and field supplies, and war. Travel had to be cancelled at the last minute due to positive COVID-19 tests. Procurement of laboratory supplies took months. The team had to deal with disinformation about the project and our collaboration. Inflated costs for supplies and travel cut into project budgets. With each hurdle thrown at the team, we resolved to stay strong and keep the focus on the science. During the first year of the project, however, in 2021–2022, team members would also ask themselves, "What the hell just happened?"

The challenges that societies now face are growing with each passing year. A never-ending pandemic and spillover infectious disease outbreaks, such as monkeypox, disrupt how we communicate and collaborate. Social distrust and the politicization of science add stresses for researchers, who now have to deal with mis- and disinformation at a grand scale. Extreme weather events and a rapidly changing environment put pressures on our resources and time. Researchers may not be able to get to their laboratory due to a forest fire or flood in the region. Or war. These issues also create supply chain difficulties and inflationary costs for laboratory products and field supplies. Borders are closed for scientists in many regions. In these difficult times, collaborations need to be stronger, and team members must be unfailing in their dedication to support each other and the research vision they have together. For it is through science that answers will come to help solve these larger issues, ranging from understanding how species will adapt (or not) to climate change to finding ways for societies to work better together (or eventually fall apart).

Now is the time for scientists to support each other and collaborate more. To up our own game and become dream teams. Now is the time for researchers to band together as communities and networks, in order to address mis- and disinformation. To stand up not only with our data and our validated experiments, but also with the stories that back up that data. Now is the time to build more diverse teams that can arrive sooner at solutions to problems that occur in projects. To tap into our diversity and ask better questions. Now is the time to move from simple inclusiveness to the full integration of our team members and their ideas. Now is the time to move forward into the future more confidently and positively together as scientists, jointly tackling the tougher questions.

REFERENCES

INTRODUCTION

Couzin-Frankel, J. 2018. Journals under the microscope. Science 361:1180–1183. https://doi.org/10.1126/science.361.6408.1180

Covey, S. M. R. 2008. The Speed of Trust: The One Thing That Changes Everything. Simon & Schuster, London, UK.

Enserink, M. 2018. Research on research. Science 361:1178–1179. https://doi.org/10.1126/science.361.6408.1178

CHAPTER 1. TRANSFORMATIVE COLLABORATIONS

Arzberger, P., P. Schroeder, A. Beaulieu, G. Bowker, K. Casey, L. Laaksonen, D. Moorman, et al. 2004. An international framework to promote access to data. Science 303:1777–1778. https://doi.org/10.1126/science.1095958

Axelrod, R., and W. D. Hamilton. 1981. The evolution of cooperation. Science 211:1390–1396. https://doi.org/10.1126/science.7466396

Bennis, P. E., and W. G. Slater. 1968. The Temporary Society. Harper & Row, New York, NY.

CNN. CNN Time Capsule: 100 Defining Moments of the Year. 1993. CNN / Vicarious Entertainment.

Deininger A., A. H. Martin, J. C. F. Pardo, P. R. Berg, J. Bhardwaj, D. Catarino, A. Fernández-Chacón, et al. 2021. Coastal research seen through an early career lens—a perspective on barriers to interdisciplinarity in Norway. Frontiers in Marine Science 8. https://doi.org/10.3389/fmars.2021.634999

Handy, C. 1995. The Age of Paradox. Harvard Business Review Press, Boston, MA.

Kuhn, T. S. 1962. The Structure of Scientific Revolutions. University of Chicago Press, Chicago, IL.

Leahey, E. 2006. Gender differences in productivity: Research specialization as a missing link. Gender & Society 20:754–780. https://doi.org/10.1177/0891243206293030

Maslow, A. H. 1943. A theory of human motivation. Psychological Review 50:370–396.

National Science Board. 2007. Enhancing Support of Transformative Research at the National Science Foundation. National Science Foundation, Arlington, VA. https://www.nsf.gov/nsb/publications/landing/nsb0732.jsp

Pannell, J. L., A. M. Dencer-Brown, S. S. Greening, E. A. Hume, R. M. Jarvis, C. Mathieu, J. Mugford, and R. Runghen. 2019. An early-career perspective on encouraging collaborative and interdisciplinary research in ecology. Ecosphere 10. https://doi.org/10.1002/ecs2.2899

Pennington, D. D., G. L. Simpson, M. S. McConnell, J. M. Fair, and R. J. Baker. 2013. Transdisciplinary research, transformative learning, and transformative science. BioScience 63:564–573. https://academic.oup.com/bioscience/article/63/7/564/289183

Senge, P. 1990. The Fifth Discipline: The Art and Practice of the Learning Organization. Doubleday/Currency, New York, NY.

Snow, N. 2005. Successfully curating smaller herbaria and natural history collections in academic settings. BioScience 559:771–779. https://academic.oup.com/bioscience/article/55/9/771/286113

Toffler, A. 1970. Future Shock. Random House, New York, NY.

Williams, G. C. 1966. Adaptation and Natural Selection. Princeton University Press, Princeton, NJ.

Zeitz, P. S., J. C. Butler, J. E. Cheek, M. C. Samuel, J. E. Childs, L. A. Shands, R. E. Turner, et al. 1995. A case-control study of hantavirus pulmonary syndrome during an outbreak in the southwestern United States. Journal of Infectious Diseases 171:864–870. https://doi.org/10.1093/infdis/171.4.864

CHAPTER 2. COMMUNITIES

AAAS.org staff report. 2009. AAAS Reaffirms Statements on Climate Change and Integrity. https://www.aaas.org/news/aaas-reaffirms-statements-climate-change-and-integrity

Asplund, M., and C. G. Welle. 2018. Advancing science: How bias holds us back. Neuron 99:635–639. https://doi.org/10.1016/j.neuron.2018.07.045

Atkinson, M., P. Doherty, and K. Kinder. 2005. Multi-agency working: Models, challenges and key factors for success. Journal of Early Childhood Research 3:7–17. https://doi.org/10.1177/1476718X05051344

Atkinson, M., A. Wilkin, A. Stott, P. Doherty, and K. Kinder. 2002. Multi-Agency Working: A Detailed Study. National Foundation for Educational Research, Slough. Berkshire, UK. https://www.nfer.ac.uk/publications/CSS02/CSS02.pdf

Bendels, M. H. K., R. Müller, D. Brueggmann, and D. A. Groneberg. 2018. Gender disparities in high-quality research revealed by Nature Index journals. PLoS ONE 13:e0189136. https://doi.org/10.1371/journal.pone.0189136

Bowles, S., and H. Gintis. 2011. A Cooperative Species: Human Reciprocity and Its Evolution. Princeton University Press, Princeton, NJ.

Bozeman, B., and E. Corley. 2004. Scientists' collaboration strategies: Implications for scientific and technical human capital. Research Policy 33:599–616. https://doi.org/10.1016/j.respol.2004.01.008

Bozeman, B., and M. Gaughan. 2011. How do men and women differ in research collaborations? An analysis of the collaborative motives and strategies of academic researchers. Research Policy 40:1393–1402. https://doi.org/10.1016/j.respol.2011.07.002

Daraghmi, E. Y., and S.-M. Yuan. 2014. We are so close, less than 4 degrees separating you and me! Computers in Human Behavior 30:273–285. https://doi.org/10.1016/j.chb.2013.09.014

DeFilippis, E., S. Impink, M. Singell, J. T. Polzer, and R. Sadun. 2020. Collaborating During Coronavirus: The Impact of COVID-19 on the Nature of Work. Harvard Business School Organizational Behavior Unit. Working Paper No. 21-006. https://papers.ssrn.com/sol3/papers.cfm?abstract_id=3654470

Engebøa, A., O. J. Klakegga, J. Lohnea, R. A. Bohnea, H. Fyhn, and O. Lædrea. 2022. High-performance building projects: How to build trust in the team. Architectural Engineering and Design Management 18:774–790.

Freeman, R. B., and W. Huang. 2015. Collaborating with people like me: Ethnic coauthorship within the United States. Journal of Labor Economics 33:S289–S318. https://doi.org/10.1086/678973

Guan, J., Y. Yan, and J. Zhang. 2015. How do collaborative features affect scientific output? Evidences from wind power field. Scientometrics 102:333–355. https://doi.org/10.1007/s11192-014-1311-x

Holman, L., D. Stuart-Fox, and C. E. Hauser. 2018. The gender gap in science: How long until women are equally represented? PloS Biology 16:e2004956. https://doi.org/10.1371/journal.pbio.2004956

Huber, J., S. Inoua, R. Kerschbamer, C. König-Kersting, S. Palan, and V. L. Smith. 2022. Nobel and Novice: Author Prominence Affects Peer Review. Working Paper No. 2022-01. School of Business, Economics and Social Sciences, Karl-Franzens-Universität Graz. Available at SSRN. https://dx.doi.org/10.2139/ssrn.4190976

Huxham, C., and S. Vangen. 2004. Realizing the advantage or succumbing to inertia? Organizational Dynamics 33:190–201. https://doi.org/10.1016/j.orgdyn.2004.01.006

King, M. M., C. T. Bergstrom, S. J. Correll, J. Jacquet, and J. D. West. 2017. Men set their own cites high: Gender and self-citation across fields and over time. Socius: Sociological Research for a Dynamic World 3. https://doi.org/10.1177/2378023117738903

Knight Foundation. 2010. Soul of the Community Survey. https://knightfoundation.org/sotc

Kyvik, S., and M. Teigen. 1996. Child care, research collaboration, and gender differences in scientific productivity. Science, Technology, & Human Values 21:54–71. https://doi.org/10.1177/016224399602100103

Lee, S., and B. Bozeman. 2005. The impact of research collaboration on scientific productivity. Social Studies of Science 35:673–702. https://doi.org/10.1177/0306312705052359

Levsky, M. E., A. Rosin, T. P. Coon, W. L. Enslow, and M. A. Miller. 2007. A descriptive analysis of authorship within medical journals, 1995–2005.

Southern Medical Journal 100:371–375. https://doi.org/10.1097/01.smj.0000257537.51929.4b

Lincoln, A. E., S. Pincus, J. B. Koster, and P. S. Leboy. 2012. The Matilda effect in science: Awards and prizes in the US, 1990s and 2000s. Social Studies of Science 42:307–320. https://doi.org/10.1177/0306312711435830

Lincoln, A. E., S. Pincus, and V. Schick. 2009. Evaluating science or evaluating gender? Back Pages section, APS News 18. https://www.aps.org/publications/apsnews/200906/backpage.cfm

McMillan, D. W., and D. M. Chavis. 1986. Sense of community: A definition and theory. Journal of Community Psychology 14:6–23. https://doi.org/10.1002/1520-6629(198601)14:1%3C6::AID-JCOP2290140103%3E3.0.CO;2-I

Merton, R. K. 1968. The Matthew effect in science: The reward and communication systems of science are considered. Science 159:56–63. https://www.science.org/doi/10.1126/science.159.3810.56

Newman, M. E. J. 2001. The structure of scientific collaboration networks. PNAS 98:404. https://doi.org/10.1073/pnas.98.2.404

Newman, M. E. J. 2004. Coauthorship networks and patterns of scientific collaboration. PNAS 101:5200–5205. https://doi.org/10.1073/pnas.0307545100

Pillemer, J., and N. P. Rothbard. 2018. Friends without benefits: Understanding the dark sides of workplace friendship. Academy of Management Review 43:635–660. https://psycnet.apa.org/doi/10.5465/amr.2016.0309

Regehr, E. V., K. L. Laidre, H. R. Akçakaya, S. C. Amstrup, T. C. Atwood, N. J. Lunn, M. Obbard, H. Stern, G. W. Thiemann, and Ø. Wiig. 2016. Conservation status of polar bears (*Ursus maritimus*) in relation to projected sea-ice declines. Biology Letters 12. https://doi.org/10.1098/rsbl.2016.0556

Rennie, D., and A. Flanagin. 1994. Authorship! authorship! Guests, ghosts, grafters, and the two-sided coin. JAMA 271:469–471. https://doi.org/10.1001/jama.1994.03510300075043

Rossiter, M. W. 1993. The Matthew Matilda effect in science. Social Studies of Science 23:325–341. https://doi.org/10.1177/030631293023002004

Schelling, T. C. 1971. Dynamic models of segregation. Journal of Mathematical Sociology 1:143–186. https://doi.org/10.1080/0022250X.1971.9989794

Sonnenwald, D. H. 2008. Scientific collaboration. Annual Review of Information Science and Technology 41:643–681. https://doi.org/10.1002/aris.2007.1440410121

Vogl, C. H. 2016. The Art of the Community: Seven Principles for Belonging. Berrett-Koehler, Oakland, CA.

CHAPTER 3. A SCIENTIFIC REVOLUTION

Aad, G., B. Abbott, J. Abdallah, O. Abdinov, R. Aban, M. Abolins, O. S. Abouzeid, et al. 2015. Combined measurement of the Higgs boson mass in *pp* collisions at $\sqrt{s} = 7$ and 8 TeV with the ATLAS and CMS experiments. Physical Review Letters 114. https://link.aps.org/doi/10.1103/PhysRevLett.114.191803

Almeida-Souza, L., and L. O'Brien. 2022. A kinder approach to science. Trends in Cell Biology 32:177–178. https://doi.org/10.1016/j.tcb.2021.11.003

Asplund, M., and C. G. Welle. 2018. Advancing science: How bias holds us back. Neuron 99:635–639. https://doi.org/10.1016/j.neuron.2018.07.045

Barber, K. 2019. Twitter post. https://twitter.com/katelizabee/status/1179869396757614609

Bendels, M. H. K., R. Müller, D. Brueggmann, and D. A. Groneberg. 2018. Gender disparities in high-quality research revealed by Nature Index journals. PLoS ONE 13:e0189136. https://doi.org/10.1371/journal.pone.0189136

Bennett, L. M., and H. Gadlin. 2019. Conflict prevention and management in science teams. Pp. 295–302 in K. Hall, A. Vogel, and R. Croyle, eds. Strategies for Team Science Success. Springer, Cham, Switzerland. https://doi.org/10.1007/978-3-030-20992-6_22

Bozeman, B., and J. Youtie. 2017. The Strength in Numbers: The New Science of Team Science. Princeton University Press, Princeton, NJ.

DeFilippis, E., S. Impink, M. Singell, J. T. Polzer, and R. Sadun. 2020. Collaborating During Coronavirus: The Impact of COVID-19 on the Nature of Work. Harvard Business School Organizational Behavior Unit Working Paper No. 21-006. https://papers.ssrn.com/sol3/papers.cfm?abstract_id=3654470

Dhawan, E. 2021. Digital Body Language: How to Build Trust and Connection, No Matter the Distance. St. Martin's, New York, NY.

Dhawan, E., and J. Saj-Nicole. 2015. Get Big Things Done: The Power of Connectional Intelligence. St. Martin's, New York, NY.

Ferrazzi, K., and N. Weyrich. 2020. Leading Without Authority: How the New Power of Co-Elevation Can Break Down Silos, Transform Teams, and Reinvent Collaboration. Currency, New York, NY.

Freeman, R. B., and W. Huang. 2015. Collaborating with people like me: Ethnic coauthorship within the United States. Journal of Labor Economics 33: S289–S318. https://doi.org/10.1086/678973

Gaughan, M., and B. Bozeman. 2016. Using the prisms of gender and rank to interpret research collaboration power dynamics. Social Studies of Science 46:536–558. https://doi.org/10.1177/0306312716652249

Holman, L., D. Stuart-Fox, and C. E. Hauser. 2018. The gender gap in science: How long until women are equally represented? PLoS Biology 16:e2004956. https://doi.org/10.1371/journal.pbio.2004956

Huber, J., S. Inoua, R. Kerschbamer, C. König-Kersting, S. Palan, and V. L. Smith. 2022. Nobel and Novice: Author Prominence Affects Peer Review. Working Paper No. 2022-01. School of Business, Economics and Social Sciences, Karl-Franzens-Universität Graz. Available at SSRN. https://dx.doi.org/10.2139/ssrn.4190976

Keating, E., and S. L. L. Jarvenpaa. 2016. Words Matter: Communicating Effectively in the New Global Office. University of California Press, Oakland, CA.

King, M. M., C. T. Bergstrom, S. J. Correll, J. Jacquet, and J. D. West. 2017. Men set their own cites high: Gender and self-citation across fields and over time.

Socius: Sociological Research for a Dynamic World 3. https://doi.org/10.1177/2378023117738903
Leahey, E. 2006. Gender differences in productivity: Research specialization as a missing link. Gender & Society 20:754–780. https://doi.org/10.1177/0891243206293030
Love, H. B., A. Stephens, B. K. Fosdick, E. Tofany, and E. R. Fisher. 2022. The impact of gender diversity on scientific research teams: A need to broaden and accelerate future research. Humanities and Social Science Communications 9:article 386. https://doi.org/10.1057/s41599-022-01389-w
Merton, R. K. 1968. The Matthew effect in science: The reward and communication systems of science are considered. Science 159:56–63. https://www.science.org/doi/10.1126/science.159.3810.56
Montaigne, M. de. 1993. Michel de Montaigne: The Complete Essays, reprint edition. Translated by M. A. Screech. Penguin Classics, London, UK.
National Center for Science and Engineering Statistics. 2019. Women, Minorities, and Persons with Disabilities in Science and Engineering. NSF-19-304. National Science Foundation, Alexandria, VA. https://ncses.nsf.gov/pubs/nsf19304
National Research Council. 2015. Enhancing the Effectiveness of Team Science. N. J. Cooke and M. L. Hilton, eds., Committee on the Science of Team Science, Board on Behavioral, Cognitive, and Sensory Sciences, Division of Behavioral and Social Sciences and Education. National Academies Press, Washington, DC.
National Science Board. 2018. Immigration and the S&E workforce. Chapter 3 in Science and Engineering Indicators. NSB-2018, Digest NSB-2018-2. https://www.nsf.gov/statistics/2018/nsb20181/report/sections/science-and-engineering-labor-force/immigration-and-the-s-e-workforce
Project Implicit. https://implicit.harvard.edu/implicit/
Rossiter, M. W. 1993. The Matthew Matilda effect in science. Social Studies of Science 23:325–341. https://doi.org/10.1177/030631293023002004
Salovy, P., and J. D. Mayer. 1990. Emotional intelligence. Imagination, Cognition, and Personality 9:185–211. https://journals.sagepub.com/doi/10.2190/DUGG-P24E-52WK-6CDG
Schelling, T. C. 1971. Dynamic models of segregation. Journal of Mathematical Sociology 1:143–186. https://doi.org/10.1080/0022250X.1971.9989794
Taylor, S. R. 2020. Instagram post. https://www.instagram.com/p/B-fc3ejAlvd/?hl=en
Yang, Y., T. Y. Tian, T. K. Woodruff, B. F. Jones, and B. Uzzi. 2022. Gender-diverse teams produce more novel and higher-impact scientific ideas. PNAS 119. https://doi.org/doi:10.1073/pnas.2200841119

CHAPTER 4. THE SCIENCE OF TEAM SCIENCE

Bozeman, B., and E. Corley. 2004. Scientists' collaboration strategies: Implications for scientific and technical human capital. Research Policy 33:599–616. https://doi.org/10.1016/j.respol.2004.01.008

Bozeman, B., and M. M. Fellows. 1988. Technology transfer at the U.S. national laboratories: A framework for evaluation. Evaluation and Program Planning 11:65–75.

Bozeman, B., and M. Gaughan. 2007. Impacts of grants and contracts on academic researchers' interactions with industry. Research Policy 36:694–707. https://doi.org/10.1016/j.respol.2007.01.007

Bozeman, B., and M. Gaughan. 2011. How do men and women differ in research collaborations? An analysis of the collaborative motives and strategies of academic researchers. Research Policy 40:1393–1402. https://doi.org/10.1016/j.respol.2011.07.002

Bozeman, B., and J. Youtie. 2017. The Strength in Numbers: The New Science of Team Science. Princeton University Press, Princeton, NJ.

Cross, R., R. Rebele, and A. Grant. 2016. Collaborative overload. Harvard Business Review, January–February. https://hbr.org/2016/01/collaborative-overload

Edmondson, A. 1999. Psychological safety and learning behavior in work teams. Administrative Science Quarterly 44:350–383. https://doi.org/10.2307/2666999

Ferrazzi, K., and N. Weyrich. 2020. Leading Without Authority: How the New Power of Co-Elevation Can Break Down Silos, Transform Teams, and Reinvent Collaboration. Currency, New York, NY.

Grant, A., and S. Sandberg. 2015. Madam C.E.O., get me a coffee. 2015. New York Times, February 6.

Grant, A. M. 2008. The significance of task significance: Job performance effects, relational mechanisms, and boundary conditions. Journal of Applied Psychology 93:108–124. https://psycnet.apa.org/doi/10.1037/0021-9010.93.1.108

Grossman, R., K. Nolan, Z. Rosch, D. Mazer, and E. Salas. 2022. The team cohesion-performance relationship: A meta-analysis exploring measurement approaches and the changing team landscape. Organizational Psychology Review 12:181–238. https://doi.org/10.1177/20413866211041157

Hall, K. L., A. L. Vogel, G. C. Huang, K. J. Serrano, E. L. Rice, S. P. Tsakraklides, and S. M. Fiore. 2018. The science of team science: A review of the empirical evidence and research gaps on collaboration in science. American Psychologist 73:532–548. https://doi.org/10.1037/amp0000319

Heilman, M. E., and J. J. Chen. Same behavior, different consequences: Reactions to men's and women's altruistic citizenship behavior. Journal of Applied Psychology 90:431–441. https://doi.org/10.1037/0021-9010.90.3.431

Hein, P., T. L. Hughes, and S. K. Golant. 2015. Hardball for Women: Winning at the Game of Business, revised and updated 3rd edition. Plume / Penguin Group, New York, NY.

Holman, L., D. Stuart-Fox, and C. E. Hauser. 2018. The gender gap in science: How long until women are equally represented? PloS Biology 16:e2004956. https://doi.org/10.1371/journal.pbio.2004956

Huxham, C., and S. Vangen. 2004. Doing things collaboratively: Realizing the advantage or succumbing to inertia? IEEE Engineering Management Review 32:11–20. https://doi.org/doi:10.1109/EMR.2004.25132

International Network for the Science of Team Science. https://www.inscits.org/about-us

Jennings, B., and J. McRandle. 2011. Attacking the lion: A study of cohesion in naval special warfare operational units. Master's thesis, Naval Postgraduate School, Monterey, CA. https://apps.dtic.mil/sti/pdfs/ADA547824.pdf

Kim, A., Y. Kim, and Y. Cho. 2023. The consequences of collaborative overload: A long-term investigation of helping behavior. Journal of Business Research 154. https://doi.org/10.1016/j.jbusres.2022.113348

Newman, M. E. J. 2001. The structure of scientific collaboration networks. PNAS 98:404–409. https://doi.org/10.1073/pnas.98.2.404

Newman, M. E. J. 2004. Coauthorship networks and patterns of scientific collaboration. PNAS 101:5200–5205. https://doi.org/10.1073/pnas.0307545100

Quast, L. 2015. Ending gender bias: Why Richard Branson says everyone should take meeting notes, not just women. Forbes, August 31.

Roghanizad, M. M., and V. K. Bohns. 2017. Ask in person: You're less persuasive than you think over email. Journal of Experimental Social Psychology 69:223–226. https://doi.org/10.1016/j.jesp.2016.10.002

Sonnenwald, D. H. 2008. Scientific collaboration. Annual Review of Information Science and Technology 41:643–681. https://doi.org/10.1002/aris.2007.1440410121

Yegros-Yegros, A., I. Rafols, and P. D'Este. 2015. Does interdisciplinary research lead to higher citation impact? The different effect of proximal and distal interdisciplinarity. PloS ONE 10. https://doi.org/10.1371/journal.pone.0135095

CHAPTER 5. TRUST

Cloud, H. 2006. Integrity: The Courage to Meet the Demands of Reality. Harper Collins, New York, NY.

Covey, S. M. R. 2008. The Speed of Trust: The One Thing That Changes Everything. Free Press, New York, NY.

Edmondson, A. 1999. Psychological safety and learning behavior in work teams. Administrative Science Quarterly 44:350–383. https://doi.org/10.2307/2666999

Fagerlin, R. 2013. Trustology: The Art and Science of Leading High-Trust Teams. Wise Guys Press, Fort Collins, CO.

Goethe, J. W. von. 2002. The Sorrows of Young Werther, Dover Thrift edition. Translated by N. H. Dole. Dover, Mineola, NY.

Grenny, J., K. Patterson, R. McMillan, A. Switzler, and E. Gregory. 2021. Crucial Conversations: Tools for Talking When Stakes Are High, 3rd edition. McGraw Hill, New York, NY.

Hull, D. 1990. Science as a Process: An Evolutionary Account of the Social and Conceptual Development of Science. University of Chicago Press, Chicago, IL.

Jargon File, version 2.1.1 (draft). 1990. http://catb.org/jargon/oldversions/jarg211.txt

Knapp, J., J. Zeratsky, and B. Kowitz. 2016. Sprint: How to Solve Big Problems and Test New Ideas in Just Five Days. Simon & Schuster, New York, NY.

Lencioni, P. 2002. The Five Dysfunctions of a Team: A Leadership Fable. Jossey-Bass, San Francisco, CA.

Rozovsky, J. 2015. The five keys to a successful Google team. Re:Work. https://rework.withgoogle.com/blog/five-keys-to-a-successful-google-team

Scrum.org. https://www.scrum.org

Shrum, W., I. Chompalov, and J. Genuth. 2001. Trust, conflict and performance in scientific collaborations. Social Studies of Science. 31:681–730. https://doi.org/10.1177/030631201031005002

Sutherland, J. J. 2014. Scrum: The Art of Doing Twice the Work in Half the Time. Crown Business, New York, NY.

CHAPTER 6. COMPETENCE

Covey, S. M. R. 2008. The Speed of Trust: The One Thing That Changes Everything. Free Press, New York, NY.

Horvitz, B. 2014. Under Armour has over-the-top 2014. USA Today, December 22, updated December 28.

Key Competence Network on School Education. 2006. Mathematical competence and basic competences in science and technology: As defined in the recommendation of the European Parliament and of the Council of 18 December 2006 on Key Competences for Lifelong Learning. http://keyconet.eun.org/maths-science-tech

Klebesadel, R. 2012. The discovery of the gamma-ray burst phenomenon. Pp.1–8 in C. Kouveliotou, R. Wijers, and S. Woosley, eds. Gamma-Ray Bursts. Cambridge Astrophysics Series. Cambridge University Press, Cambridge, UK. https://doi.org/10.1017/CBO9780511980336.002

Klebesadel, R., I. B. Strong, and R. A. Olson. 1973. Observations of gamma-ray bursts of cosmic origin. Letters section, Astrophysical Journal 182:L85-L88.

Plait, Phil. 2022. The brightest gamma-ray burst ever recorded rattled Earth's atmosphere. Scientific American, Astronomy—Opinion section, October 21. https://www.scientificamerican.com/article/the-brightest-gamma-ray-burst-ever-recorded-rattled-earths-atmosphere

Shilling, G. 2002. Flash! The Hunt for the Biggest Explosions in the Universe. Cambridge University Press, Cambridge, UK.

CHAPTER 7. COMMUNICATION

Rutledge, R. B., N. Skandali, P. Dayan, and R. J. Dolan. 2014. A computational and neural model of momentary subjective well-being. PNAS 111:12252–12257. https://doi.org/10.1073/pnas.1407535111

CHAPTER 8. FISH DON'T KNOW THEY'RE IN WATER

Christov-Moore, L., E. A. Simpson, G. Coudé, K. Grigaityte, M. Lacoboni, and P. F. Ferrari. 2014. Empathy: Gender effects in brain and behavior. Neuroscience & Biobehavioral Reviews 46:604–627. https://doi.org/10.1016/j.neubiorev.2014.09.001

Dweck, C. S. 2017. Mindset: Changing the Way You Think to Fufill Your Potential, revised edition. Robinson, London, UK.

Grant, A. M. 2008. The significance of task significance: Job performance effects, relational mechanisms, and boundary conditions. Journal of Applied Psychology 93:108–124. https://doi.org/10.1037/0021-9010.93.1.108

Klein, K. J. K., and S. D. Hodges. 2001. Gender differences, motivation, and empathic accuracy: When it pays to understand. Personality and Social Psychology Bulletin 27:720–730.

Konrath, S. H., E. H. O'Brien, and C. Hsing. 2011. Changes in dispositional empathy in American college students over time: A meta-analysis. Personality and Social Psychology Review 15:180–198. https://doi.org/10.1177/1088868310377395

Schumann, K., J. Zaki, and C. S. Dweck. 2014. Addressing the empathy deficit: Beliefs about the malleability of empathy predict effortful responses when empathy is challenging. Journal of Personality and Social Psychology 107:475–493. https://doi.org/10.1037/a0036738

Seidman, D. 2007. How: Why How We Do Anything Means Everything . . . in Business and in Life, expanded edition. John Wiley & Sons, Hoboken, NJ.

Zaki, J. 2019. The War for Kindness: Building Empathy in a Fractured World. Crown, New York, NY.

CHAPTER 9. DREAM TEAMS

Bozeman, B., and E. Corley. 2004. Scientists' collaboration strategies: Implications for scientific and technical human capital. Research Policy 33:599–616.

Bozeman, B., and M. Gaughan. 2011. How do men and women differ in research collaborations? An analysis of the collaborative motives and strategies of academic researchers. Research Policy 40:1393–1402. https://doi.org/10.1016/j.respol.2011.07.002

Bozeman, B., and J. Youtie. 2017. The Strength in Numbers: The New Science of Team Science. Princeton University Press, Princeton, NJ.

Covey, S. M. R. 2008. The Speed of Trust: The One Thing That Changes Everything. Simon & Schuster, London, UK.

Grenny, J., K. Patterson, R. McMillan, A. Switzler, and E. Gregory. 2021. Crucial Conversations: Tools for Talking When Stakes Are High, 3rd edition. McGraw Hill, New York, NY.

Lee, S., and B. Bozeman. 2005. The impact of research collaboration on scientific productivity. Social Studies of Science 35:673–702. https://doi.org/10.1177/0306312705052359

Maxwell, J. C. 2005. The 360° Leader: Developing Your Influence from Anywhere in the Organization. Nelson Business, Nashville, TN.

Maxwell, J. C. 2018. Executive leadership podcast #5: 3 questions every follower is asking about their leader. May 15. https://corporatesolutions.johnmaxwell.com/podcast/executive-leadership-podcast-5–3-questions-every-follower-is

Methot, J. R., J. A. Lepine, N. P. Podsakoff, and J. S. Christian. 2016. Are workplace friendships a mixed blessing? Exploring tradeoffs of multiplex relationships and their associations with job performance. Personnel Psychology 69:311–355. https://doi.org/10.1111/peps.12109

Olivet Nazarene University. 2019. Research on Friends at Work. https://online.olivet.edu/news/research-friends-work

Page, S. 2017. The Diversity Bonus: How Great Teams Pay Off in the Knowledge Economy. Princeton University Press, Princeton, NJ.

Senge, P. 1990. The Fifth Discipline: The Art and Practice of the Learning Organization. Doubleday/Currency, New York, NY.

Zani, M. 2021. The Science of Dream Teams: How Talent Optimization Can Drive Engagement, Productivity, and Happiness. McGraw Hill, New York, NY.

CHAPTER 10. SCIENCE NETWORKS

Angere, S., and E. J. Olsson. 2017. Publish late, publish rarely! Network density and group performance in scientific communication. Pp. 34–62 in T. Boyer-Kassem, C. Mayo-Wilson, and M. Weisberg, eds. Scientific Collaboration and Collective Knowledge: New Essays. Oxford University Press, NY. https://doi.org/10.1093/oso/9780190680534.003.0002

Hahn, U., J. U. Hansen, and E. J. Olsson. 2020. Truth tracking performance of social networks: How connectivity and clustering can make groups less competent. Synthese 197:1511–1541. https://doi.org/10.1007/s11229-018-01936-6

National Science Foundation. Research Coordination Networks. https://beta.nsf.gov/funding/opportunities/research-coordination-networks

Olsson, E. J. 2005. Against Coherence: Truth, Probability, and Justification. Oxford University Press, Oxford, UK.

Olsson, E. J. 2020. Why Bayesian agents polarize. Chapter 11 in F. Broncano-Berrocal and A. Carter, eds. The Epistemology of Group Disagreement. Routledge, NY. https://lucris.lub.lu.se/ws/portalfiles/portal/78949025/Why_Bayesian_agents_polarize.pdf

Olsson, E. J. 2022. Coherentism: Elements in Epistemology. Cambridge University Press, Cambridge, UK.

Research Coordination Networks (RCN). Program announcements and information. Posted December 14, 2022. https://www.nsf.gov/publications/pub_summ.jsp?ods_key=nsf23529

Vallinder, A., and E. J. Olsson. 2014. Trust and the value of overconfidence: A Bayesian perspective on social network communication. Synthese 191:1991–2007. https://doi.org/10.1007/s11229-013-0375-0

CHAPTER 11. WHAT THE HELL JUST HAPPENED?

Almeida-Souza, L., and L. O'Brien, 2022. A kinder approach to science. Trends in Cell Biology 32:177–178. https://doi.org/10.1016/j.tcb.2021.11.003

INDEX